Public Deliberation on Climate Change

Public Deliberation on Climate Change

Lessons from Alberta Climate Dialogue

EDITED BY LORELEI L. HANSON

AU PRESS

Copyright © 2018 Lorelei L. Hanson

Published by AU Press, Athabasca University
1200, 10011 – 109 Street, Edmonton, AB T5J 3S8

ISBN 978-1-77199-215-2 (pbk.)   978-1-77199-216-9 (PDF)   978-1-77199-217-6 (epub)
doi: 10.15215/aupress/9781771992152.01

Cover design by Marvin Harder
Interior design by Sergiy Kozakov
Illustrations by Liisa Sorsa
Printed and bound in Canada by Friesens

Library and Archives Canada Cataloguing in Publication

Public deliberation on climate change: lessons from Alberta Climate Dialogue /
edited by Lorelei L. Hanson.

Includes bibliographical references and index.
Issued in print and electronic formats.

1. Climatic changes—Government policy—Alberta—Citizen participation. 2. Climatic
changes—Social aspects—Alberta. 3. Discussion. I. Hanson, Lorelei Lynn, 1964–, editor

TD171.7.P83 2018   363.7'0525   C2017-907386-9
                                              C2017-907387-7

This book has been published with the help of a grant from the Federation for the
Humanities and Social Sciences, through the Awards to Scholarly Publications Program,
using funds provided by the Social Sciences and Humanities Research Council of Canada.

We acknowledge the financial support of the Government of Canada through the
Canada Book Fund (CBF) for our publishing activities and the assistance provided by the
Government of Alberta through the Alberta Media Fund.

This book is dedicated to the hundreds of citizens who so willingly participated in the dialogues, without whom this book, and the research upon which it is based, would not have been possible. We are grateful for the commitment and thoughtful reflection that you brought to the dialogue processes.

Contents

Acknowledgements

This book was a collaborative effort in many ways and would not have been possible without the support, participation, and contribution of many, including Vincent Ambrock, Shelley Boulianne, Jacquie Dale, Susanna Haas Lyons, Janette Hartz-Karp, Mikael Hellstrom, Simon Knight, Matt Leighninger, Kristjana Loptson, Mary Pat MacKinnon, Melanie Marvin, John Parkins, Tom Prugh, Alex Ryan, Geoff Salomons, and participants in the 2014 and 2016 NOMIS Workshops on Nature and Value. For constant support and guidance throughout the development of the book, and for his courage, determination, and inspiration to initiate and lead a complex project like Alberta Climate Dialogue (ABCD), a huge thanks to David Kahane. For helping to formulate the framework of the book, and providing project vision and critical feedback, thanks to Gwendolyn Blue. For capturing the work of ABCD in such a beautiful and expressive visual form, thanks to Liisa Sorsa. For her generous assistance in multiple ways, thanks to Deb Schrader. For those who agreed to collaborate with us, and experiment on engaging citizens using public deliberation, appreciation is extended to: the City of Edmonton, especially Mayor Don Iveson and the Office of the Environment (particularly Jim Andrais and Mark Brostrom); Fiona Cavanagh and the Centre for Public Involvement; Jesse Row and the Alberta Energy Efficiency Alliance; and Shannon Frank and the Oldman Watershed Council. I am also grateful for the funding we received from the Social Sciences and Humanities Research Council of Canada, the Award to Scholarly Publications Program, Athabasca University, the University of Alberta, and the Centre for Public Involvement which made the research on public deliberation and this book possible. Finally, thank you to those friends and loved ones who provided encouragement and support over the course of the past year, including Hollee Card, Lynnette Kaminiski, Deb Schrader, Meenal Shrivastava, and most importantly, Desy Sarmiento Flores.

Foreword

The issue of climate change is close to my heart. Since joining Edmonton City Council in 2007, I have been committed to bold action to mitigate greenhouse gas emissions. Addressing emissions not only relates to the broader issue of climate disruption, it also factors into many other health, economic, and quality of life benefits. If Edmontonians want a vibrant, innovative, globally competitive city, we must become a leader on climate change and energy transition.

When The Way We Green plan came before Council in 2011, much of the discussion centered on what kind of engagement we would undertake to move an aspirational catalogue of measures toward implementation. I was enthusiastic when Alberta Climate Dialogue (ABCD) approached me that same year about building a citizen-deliberation process into City decision-making on climate and energy questions. As the Council lead on the Environment portfolio, I reminded my colleagues that on the really tough aspects of climate change we had a rare opportunity to connect with global networks to help a variety of decision-making bodies navigate these problems.

Something municipalities have done effectively in Canada, and around the world, is to lead when there is an absence of leadership from sub-national and national governments. Edmonton City Council understood that a municipality could not alone deliver solutions to emissions for our city, much less our region, province, or country. That said, the majority of Edmonton councillors were comfortable with climate science which gave us an opportunity to make an important progressive statement. With regard to our responsibility for the emissions output from this community, we thought it essential to provide strong leadership where we could.

As an elected representative, I am reassured to know that when a diverse group of citizens is brought together in a room, given the facts and a chance to fully discuss an issue, that group will let you know with a remarkable degree of consensus what ought to be done. As a decision maker, I communicated to Edmonton City Council the deliberative process we were building with ABCD, which we considered a proxy for what citizens, so informed, would come up with. I described how public deliberation was a way to get beyond the issue polarization that we had struggled against, which resulted in overly general discussions of the issues with only directly affected special interests coming forward. We knew that including all relevant interests in the public deliberation was crucial to the success of this process. After all, Edmonton is an industry town with a huge stake in the fossil fuel business. If that perspective had been excluded, we would not have had all the stakeholders in the room and the process could have easily been discredited.

However, our society is more complicated and broadly composed than special interests and industry voices might suggest. It contains people of all age groups, from different cultural, linguistic, religious, and economic backgrounds; folks with various perspectives on these vexing questions. For example, Edmonton will very soon have the largest urban Indigenous population in the country and be the fifth-largest destination for new Canadians.[1] If we do not include individuals and groups that reflect the makeup of our city, then obviously we will miss the mark. The process had to genuinely register the diverse voices of our community to have credibility with the rest of the citizenry.

I was fortunate to attend sessions of the Edmonton panel of the ABCD and had the opportunity to observe the process at work. At these sessions, I interacted with panelists and saw true diversity reflected—the kind of diversity that you would observe in a room of regular Edmontonians. I watched the journey unfold over six weeks and noted the remarkable consensus that the participants came to by the conclusion of the process. As a first-hand observer, I can say that the process had real meaning. Subsequently, I remained in contact with some of the panelists who had taken it upon themselves to advocate and lead the community on this issue. That was one of the most inspiring dividends—that and giving our

---

1 Statistics Canada. (2011). "Immigration and Ethnocultural Diversity in Canada." http://www12.statcan.gc.ca/nhs-enm/2011/as-sa/99-010-x/99-010-x2011001-eng.cfm

city administration the confidence that citizens, given the opportunity to debate and become informed about this issue, would arrive at a strong consensus, the very same consensus that our Council reached when the Energy Transition Strategy came to us for a decision in 2015.

The Citizens' Panel showed Council that a representative group of citizens, armed with good information, had come to the same conclusion that our administration recommended: that we should take action, that there was an upside to this action, and that reducing greenhouse gas impact was not the only benefit. Addressing greenhouse gases is an important issue in itself but there are associated benefits to mitigating climate change around cost savings for individuals and businesses, as well as improvements in air quality that everybody can appreciate, regardless of their position on the larger matter of reducing reliance on carbon. What surprised me most about the Edmonton deliberation is that the panelists remained engaged even though it took over two years for their recommendations to move from the panel process to Council approving the Energy Transition Strategy. The determination of the panelists to remain involved— remain champions of the strategy that they supported as citizens—is inspiring.

Given the value of the Citizens' Panel, I think that deliberative methodology could restore a kind of authenticity to citizen engagement in many other tough conversations that cities need to have. It's not right for everything we do, but it's valuable for some of the more intractable problems: whether it's poverty or economic diversification. At this moment, I do not know of any city where people feel genuinely connected to their government, think they have a say in its functioning, or see their views reflected as much as they might if there were effective deliberative processes embedded in the way the city came to decisions.

When it comes to understanding what Edmontonians think about particular and multifaceted issues like climate change, good deliberation procedures would make jobs like mine a lot easier. We can conduct surveys to figure out how citizens think based on their current knowledge of a topic, yet that is often an incomplete picture of the state of things. If citizens can see a representative group of their peers (who have gone through an authentic deliberative process) come to a set of recommendations that may be challenging and not simply compromised, middle-of-the-road outcomes, then a policymaker can conclude with reasonable confidence that citizens will accept such strong recommendations. The records from these kinds of guided and informed deliberation processes also provide open and transparent information to citizens wanting to know how a given deliberative panel arrived at their decisions. This book is an example of such a record while also containing reflections on the landscape of deliberative democracy as a movement for re-invigorating public consultations.

The City of Edmonton's partnership with ABCD produced a valuable deliberative process on climate change and energy transition. I think it is a model that we will look to again when tackling some of the other wicked problems that face cities and communities in the twenty-first century. The book in front of you contains the lessons of that deliberative engagement—though it is the strategic plan and its implementation that are ultimately its substance. Hopefully these types of activities can begin to positively change the discussion, and public perception, about the democratic role of informed consultation in city government.

Don Iveson
Mayor, City of Edmonton

Don Iveson was elected as a City Councillor for Southwest Edmonton in 2007. During his two terms as Councillor, he was the lead for the Environment portfolio when a high level, aspirational strategic plan called The Way We Green was developed and passed by Council. The most controversial element of this Strategy dealt with climate change and energy transition. Mayor Iveson recognized citizen deliberation as a way to move City work on climate and energy transition forward, and he helped Alberta Climate Dialogue develop the partnership with the City that would lead to the Citizens' Panel on Edmonton's Energy and Climate Challenges, which was convened from October to December of 2012. In 2013, Iveson was elected Mayor of Edmonton, where he saw the Citizens' Panel recommendations incorporated into an Energy Transition Strategy that was passed unanimously by City Council in May 2015.

Public Deliberation on Climate Change

Introduction

Advancing Public Deliberation on Climate Change and Other Wicked Problems

Lorelei L. Hanson and David Kahane

There is growing recognition that our current modes of public problem solving fail us in many respects, and that new methods of public exchange and decision making are required. Widespread political engagement, adaptive responses, inclusive and innovative collaborative strategies, and collective behaviour changes are needed to address the daunting challenges of our age. Traditional public decision-making processes typically do not accommodate a broad diversity of perspectives and values, or foster sufficient interaction and mutual understanding of perspectives to enable ordinary citizens to engage with the trade-offs and tensions between values. Decision-making processes more often rely on experts to determine the best solutions, or incentivize the expression of polarized positions by a select few. Public deliberation offers an alternative way of addressing our toughest public problems.

This collection examines the multiple tensions and trade-offs that emerge in deliberative citizen engagement processes addressing wicked issues like climate change. Multi-faceted and complex, wicked problems call for ongoing adaptive changes that are tailored to specific contexts and connected to multiple organizational structures and geographies. Wise responses to wicked problems such as climate change require a focus on whole systems, yet the deliberation of dozens of citizens over a few days directed at public policy and procedure recommendations makes it necessary to focus the choices and policy questions and connect them with particular local realities. This volume explores the art of balancing such tensions in deliberative engagements: looking at big picture questions and the brass tacks of particular choices, framing wicked social issues, working with diverse stakeholders and project collaborators, linking citizen processes to expert knowledge, and connecting with other social change processes and actors. It explores these balancing acts in the context of generating effective and urgent responses to a wicked issue that remains controversial and fundamentally challenging.

The four deliberation case studies that form the backdrop of the book were associated with a five-year community–university research project called Alberta Climate Dialogue (ABCD) that drew together a network of scholars, facilitation practitioners, citizens, members of civil society organizations, government officials, and not-for-profits from Canada and other parts of the English-speaking world. Members of ABCD participated in four deliberations connected in varying degrees to climate change, engaging citizens from across the Canadian province of Alberta. Throughout the deliberations, ABCD undertook research (surveys, citizen journalling, interviews, and observations) to assess the impacts of the deliberations on policy outcomes and citizen views. Many of the contributors to this volume were centrally involved with ABCD as process facilitators and/or researchers, and came to the project with a range of experiences, backgrounds, and concerns. Drawing on the work of ABCD, this book explores how to organize, convene, and evaluate public deliberation events in light of theoretically derived "big picture" questions. Core themes include:

- How to design public deliberations to engage effectively with the complexity of wicked problems
- Collaboration involved in developing a core team, building partnerships to carry out deliberations, and working with individuals and groups outside of the core team
- Forms of learning in deliberations, and about deliberations
- Using sources of knowledge that cut across expert and lay, scientific and technical claims, and interests and values
- Understanding the impacts of deliberations on public policy, participants, and broader systems

Although we struggled at times in our deliberative work on climate change in ABCD, we persevered, made mistakes, and had successes. Throughout this book we examine issues and moments that both tripped us up and resulted in innovation. ABCD was born out of a conviction that we need to build capacity within our institutions and society to find new ways to discuss and solve the complex problems of our time, and we willingly share here our experience and knowledge with you to advance public conversations and processes.

Alberta Climate Dialogue: Its History and Legacy

ABCD had its origins at a Washington, DC, meeting of the Deliberative Democracy Consortium (DDC) Researchers and Practitioners Group in 2007, a gathering of thought and practice leaders focused on collaboration to advance the field. At that meeting, about a dozen participants identified climate change as the issue that kept them up at night. Given their convictions and experiences around the usefulness of innovative citizen involvement in addressing wicked problems, they decided to work toward convening groups of citizens on climate issues to support learning and political change. This dual commitment to deliberative democracy and to supporting effective climate responses remained foundational to the project.

David Kahane, who would become ABCD's Project Director and the Principal Investigator of its research grant, suggested that the on-the-ground learning take place in Alberta. Alberta was an interesting and complicated setting in which to undertake public engagement on climate change. For decades, oil and gas have been the motor of the provincial economy; the Progressive Conservatives (a right-of-centre political party in power from 1971 until the end of ABCD's deliberations) were strongly supportive of this industry, including the Athabasca oil sands (also known as the tar sands). Successive Progressive Conservative governments had pushed back at federal environmental regulations and international conventions they viewed as threats to the oil industry, and public consultation on climate policy had been designed very cautiously and strategically (see Adkin et al. 2016).

The DDC provided seed funding for a three-day workshop that brought together cross-sectoral leaders from Alberta, including elected officials and civil servants from municipal and provincial governments, and deliberative democracy experts from around the world. While the event built enthusiasm for participatory approaches to climate responses in the province, there were struggles to sustain momentum in the year that followed, given the many different ideas about how citizens might become involved. A core group of deliberative democracy researchers and practitioners met in 2009 to develop a grant application to the Social Sciences and Humanities Research Council of Canada; they proposed to work with governments to convene citizen deliberations focused on policy, and they prioritized municipal collaborations in view of pessimism about the provincial government's willingness to partner. The application was successful, providing C\$1 million for community–university research during 2010–15, with the requirement that research accompany each citizen deliberation.

As ABCD took shape as a formal project with twenty-three community partners and thirty-five researchers, there were ongoing challenges in aligning understandings of climate change and deliberative democracy. The core team that would undertake much of the work included six seasoned researchers and practitioners of deliberative democracy, and an equal number of other ABCD members, most of them academics who did not come from the deliberative democracy world. ABCD team members struggled with how this approach could contribute to political or systemic change around climate issues. Further, while ethical and political concerns about climate change were widely shared in the project, few team members were experienced researchers or activists around the issue. The project team was further divided by institutional affiliations and expectations, and by geographic location: it included academics, most located in Alberta's capital of Edmonton but some contributing from a distance; deliberation practitioners and consultants coming mostly from outside of Alberta; and non-government Alberta organization partners, whose own organizations' approaches were not completely aligned with the vernacular of "deliberation" and associated approaches to political change. Over the years other participants were involved in specific ABCD projects and workshops, coming from energy companies, Indigenous communities and organizations, community groups, and the provincial government; many of these participants did not sustain their involvement in the project. Building common understandings was thus a constant challenge, made more difficult because many members of the team were involved in ABCD off the corners of their desks and had diverse reasons for being involved.

ABCD was challenged in its first two years to develop deliberation partnerships with government: these were vital to our aspiration to support effective climate responses, and also to the case-based research we had been funded to carry out. ABCD had strong capacity to offer around innovative citizen involvement, but we were "selling" approaches to governments and other organizations in Alberta that were not actively seeking them. Moreover,

ABCD's goal of convening citizen deliberations to support better climate responses meant that we were seeking to collaborate on participatory processes that grappled directly with climate questions, whereas some partners often were interested in help on a range of engagement projects not framed in terms of climate. While some in ABCD searched for and developed partnerships, two ABCD members observed and undertook research activities on a public deliberation convened by the Centre for Public Involvement (CPI), a small not-for-profit jointly funded by the University of Alberta and the City of Edmonton. Although that deliberation focused on the development of an urban food and agriculture strategy and was only peripherally related to climate change, it provided an opportunity for ABCD members to witness deliberation in action. After many months, a partnership was forged with the City of Edmonton in 2012 that would result in the Citizens' Panel on Edmonton's Energy and Climate Challenges. In the third year of the project, ABCD explored the possibility of a civil society-based deliberative process, but this was not judged to be promising given our capacity. Instead a call was sent to members of ABCD to propose deliberation projects, which produced two further deliberative partnerships, with the Oldman Watershed Council and the Alberta Energy Efficiency Alliance.

When it came to the models actually used in these deliberations, two factors loomed large: the goals and priorities of partners who were going to use the outcomes of deliberation in their decision making; and which practitioners in ABCD were willing not only to advocate for particular approaches but to contribute substantial time to project development. Because of the needs of partners, as well as the orientation of ABCD members and the practitioners available to help design and facilitate deliberations, ABCD's deliberations all were mini-publics, which involve diverse individuals invited to participate in carefully structured, facilitated deliberation that produces learning or recommendations for an organization that has helped to convene the event.

Over the years, many of us in ABCD gained an appreciation of each other's perspectives and approaches, but as we look back on our work we also realize that there remain a number of key differences of opinion on how best to approach public deliberation and to respond effectively to climate change. This persistent diversity of perspectives reflects the diversity of ABCD's members and their levels of involvement, the messy realities of public deliberation, and the complexity and tensions of collaboration on wicked problems.

A Super Wicked Problem

December 2015 marked what many consider a historic event: the world's first global climate agreement was reached as a part of the twenty-first Conference of the Parties to the United Nation's (UN's) Framework Convention on Climate Change (COP 21). The Paris Agreement, negotiated by the European Union and 195 nation states, expressed commitment to holding global warming to an increase of 1.5°C above pre-industrial levels, and to achieve a carbon-neutral world by 2050. The agreement also obligated industrialized nations by 2020 to collectively provide $100 billion annually to assist developing countries in adapting to the impacts of climate change and reducing greenhouse gas (GHG) emissions (UN 2015).

This historic climate change deal was met with simultaneous fanfare and criticism. *Guardian* columnist George Monbiot (2015) captured this paradox of reaction in his characterization of the Paris Agreement as both miracle and disaster. In his words:

> Inside the narrow frame within which the talks have taken place, the draft agreement at the UN climate talks in Paris is a great success . . . its aspirational limit of 1.5C of global warming, after the rejection of this demand for so many years, can be seen within this frame as a resounding victory . . . . Outside the frame it looks like something else. . . .Though negotiated by some nations in good faith, the real outcomes are likely to commit us to levels of climate breakdown that will be dangerous to all and lethal to some.

The pre-conference pledges provided by 186 countries would result in a global temperature rise of 2.7°C or higher (République Française 2015), but the agreement included a framework for the pledges to be expanded and strengthened. The 1.5°C goal was a key demand of developing countries already experiencing the harsh effects of climate change and rising sea levels, but the lack of a clear strategy to maintain temperatures at that level, the voluntary nature of the emission targets, lack of a strict timeline, and the non-binding nature of the commitments rendered the deal meaningless in the view of many (Harvey 2015; Lukacs 2015). While aspirational, the Paris Agreement is short on details about how comprehensive emission reductions and compensation to less developed nations can be practically achieved and politically enforced.

In spite of nearly universal consensus in the scientific community that there is a causal connection between climate change and human activity (IPCC 2015), governance institutions have repeatedly failed in creating policies to effectively address climate change. These failures are in part a result of the "wicked" nature of this problem. Traditional responses to policy issues work from problems to solutions: the problem is defined, outcomes and outputs determined, implementation plans designed, and performance targets specified. Yet because of their non-linear and unpredictable trajectories, wicked problems defy such approaches to problem solving. Over forty years ago, Horst Rittel and Melvin Webber (1973) detailed the features of wicked issues that challenge "rational" policy approaches like cost-benefit analysis; these social problems defy straightforward planning responses because:

- Wicked problems are difficult to clearly define—there is more than one explanation of the problem, and proposed solutions depend on how the problem is defined.

- Wicked problems involve multiple interdependencies and are often multi-causal—stakeholders have different views, interests, and goals.

- Wicked problems result in unforeseen consequences—they exist in complex systems that exhibit unpredictable, emergent behaviours.

- Wicked problems are not stable—understanding of the problem constantly evolves.

- Wicked problems have no clear solutions—solutions can only be good or bad, not true or false.

- Wicked problems are socially complex, and this, more than their technical features, makes them overwhelming.

- Wicked problems seldom are the responsibility of only one jurisdiction, organization or authority—typically these problems cross many boundaries.

- Wicked problems require changing behaviours—innovative methods beyond legislation, fines, and taxes are required to motivate organizations and individuals to actively co-operate on transformation.

- Wicked problems are characterized by chronic policy failures—they are intractable despite numerous attempts to solve them. (APSC 2007; Riedy 2013)

Many social issues are difficult to solve, but a problem is wicked if it is "indeterminate in time and scale" and therefore "can't be fixed" (Kolko 2012); and if high stakes, incomplete and contradictory information, diverse perspectives, multiple engaged interests, and continually changing contexts undermine traditional means of decision making (Hernández 2014; Kolko 2012; Lorenzoni and Pidgeon 2006). Mitigating the negative consequences of wicked problems and finding more desirable trajectories of social development require holistic rather than partial and linear thinking, innovative and flexible approaches, long-term and coordinated responses across government boundaries and between different sectors of society, and toleration of uncertainty (Collins and Ison 2009a and 2009b; Hale 2010); most existing policy and planning mechanisms and systems are incapable of effectively operating in this way.

In the case of climate change, the interdependency, circularity, and uncertainty associated with wicked problems are further confounded by a set of additionally troublesome features that make this a "super wicked" problem:

- Time is running out.
- Those causing the problem are also those proposing the solutions.
- The central authority tasked with solving the problem is weak or non-existent.
- The proposed policy responses irrationally discount the future, in part because the causes of climate change are invisible, and the impacts distant, in terms of both time and geography (Moser 2010; Levin et al. 2012).

Super wicked problems create a tragic dilemma: "even when we collectively recognize the need to act now to avoid the catastrophic impacts, the immediate implications of required behavioral changes overwhelm our collective interest in policy change and the ability of the political and policy systems at multiple levels to respond" (Levin et al. 2012, 148). Such has been the tragic response to climate change: too many times, climate agreements have become merely aspirational statements that are largely ignored after they are signed because those most responsible for and able to address this global predicament must get back to business as usual.

The repeated failures of international climate negotiations to arrive at an agreement adequate to the scale and complexity of the problem feed the conviction that more effective action on climate change is impossible. Yet there are myriad examples demonstrating humanity's remarkable ability to unmake social norms to prevent further tragedy. Glimmers of hope for humanity's ability to correct its course of action can be seen, for instance, in jurisdictions enshrining the rights of nature into law—this has happened in Ecuador, followed by Bolivia and Pittsburgh (Klein 2014)—precipitous declines in violence among humans over the last century, including decreased rates of genocide, war, homicide, sexual and domestic violence, violence against homosexuals, and capital punishment (Pinker 2015); and regeneration of bio-diversity and depleted soils through perennial, polyculture food production demonstrated by individuals and organizations in different parts of the world, including the Land Institute in Kansas (Land Institute 2017) and the Zatuna Farm in Australia (Lawton 2017). More specifically, the work of social move-ments challenging social and economic reliance on fossil fuels through court proceedings, protests, civil disobedience, and divestment campaigns (Klein 2014; Lukacs 2015) provides hopeful signs of a convergence of diverse con-stituencies around compelling action on climate change.

ABCD was driven by the conviction that bold and immediate climate action is both required and possible, and that citizens can play direct and indirect roles in bringing about such change. As many of the chapters will demon-strate, we struggled among ourselves and with our partners, and worked with citizens to address the layered complexity and "super wickedness" of climate change in a political context where the very existence of climate change was still being debated.

But What Kind of Wicked Problem Is It?

In the days leading up to the Paris Agreement, billboards were put up in Alberta's two metropolitan centres, Edmonton and Calgary, stating: "Global Warming? Not for 18+ years!" and "The sun is the main driver of climate change. Not you. Not CO2." The environmental law charity Ecojustice filed a complaint with the federal Competition Bureau asking it to investigate the false and mislead-ing claims made by the Friends of Science, the International Climate Science Coalition, and the Heartland Institute (Kent 2015, Mandel 2015), which had sponsored the billboards. According to Thomas Duck, a Canadian atmospheric scientist who also signed the complaint, "These groups attempt to discredit the established scientific consensus that global warming and climate change are real and caused by human activity. The reality, causes and consequences of climate change are well understood" (Mandel 2015).

Despite Duck's insistence that the dispute over climate change is settled, the billboards as well as the COP 21 negotiations are evidence that many aspects of this debate remain unresolved. The billboard incident highlights a contextual factor central to the case studies featured in this book: in spite of recent elections of provincial and federal governments committed to addressing climate change, climate denial, though a minority voice, still flourishes in Alberta. Yet even if everyone were to agree that the climate is changing and that these changes are largely caused by human activities, this would not resolve all questions about the biophysical uncertainties of this physical phenomenon, or the epistemological and value questions associated with it. Variable meanings circulate about what exactly climate change is and what should be done to address it (Hulme 2009a). Like all wicked issues, climate change is characterized by fierce contestation about how to define the problem and its solution.

To date, climate change has largely been framed as an urgent scientific issue. Formation of global research networks like the International Geosphere–Biosphere Program in 1986 and the International Panel on Climate Change (IPCC) in 1988 have enabled collection of large amounts of data and research from numerous scientific disciplines, expanding our knowledge about the sources, nature, and implications of climate change (Brulle and Dunlap 2015). For example, the IPCC's assessment reports (1990, 1995, 2001, 2007, and 2014) have outlined increases in anthropogenic GHG emissions and global temperatures, and the resulting environmental risks posed by these changes, including melting glaciers, warming oceans, increased precipitation, changes in extreme weather and climate events, and future projections of these changes and impacts. As well, more recent IPCC reports address both climate change mitigation (limiting GHG emissions) and, to a lesser extent, adaptation (enhancing our ability to respond to existing and emerging impacts from climate change); detail the likely socio-economic consequences of climate change with respect to food security, human health, species extinction, water availability, and displacement of people; and outline effective responses to climate mitigation and adaptation (IPCC 2014).

Each finding is accompanied by an indication of the degree of confidence in the assertion, ranging from very low to very high: for example, there is high confidence that anthropogenic activities have increased atmospheric levels of GHGs like CO_2, and low confidence that mitigation policies could raise the price for some energy services and hamper the ability to expand modern services to those most in need (IPCC 2014). In spite of the uncertainties, through the IPCC and other scientific forums, the "organized power of science" has initiated and shaped a public policy debate about the most urgent and important issue facing the planet (Szersynski and Urry 2010, 2). The message they provide is clear: "Climate change has the characteristics of a collective action problem at the global scale" with "very high risk of severe, widespread impacts globally" (IPCC 2014, 17).

In presenting not only data about global climate trends, conditions, and projections but also prescriptions for effective mitigation and adaptation strategies, climate scientists have stepped beyond the role of detached, objective observers into making value-laden judgments about not only what is dangerous but what they consider an appropriate response to this danger. In other words, "Science is being used to justify claims not merely about how the world is (what are called 'positive' statements), but about what is or is not desirable—about how the world *should* be ('normative' statements)" (Hulme 2009a, 74, emphasis in original). Given how carefully vetted the IPCC membership is, these reports could be taken to communicate the further message that only experts and elites are able to offer an accurate estimation of the climate change problem, its impacts, and legitimate responses to it (Beck 2010).

The intersection of science and politics is nothing new. For example, former US Vice President and Nobel Laureate Al Gore, in expressing support for the US Global Climate Change Research Program, declared that "more research and better research and better targeted research is [*sic*] absolutely essential if we are going to eliminate the remaining areas of uncertainty and build the broader and stronger political consensus necessary for the unprecedented actions required to address this problem" (SCCST 1989, in Pielke 2007, 87). Climate scientist Mike Hulme, who worked at the University of East Anglia's Climate Research Unit in 2009 when the institute's professional emails were stolen and used to seed public doubt about the existence of climate change, publicly reflected on the "Climategate" controversy by highlighting how climate science was being used as a proxy for political battles. In his words, the mantra "Get the science right, reduce the scientific uncertainties, compel everyone to believe it . . . and we will have won . . . is [not] only an unrealistic view about how policy gets made, it also places much too great a burden on science, certainly climate science" (Hulme 2009b). Many of the issues associated with mitigation of and adaptation to climate change are highly technical and complex. Scientific information is clearly needed, but science has its limits as well. In decision contexts where there are shared commitments to a specific goal and little uncertainty, it is quite possible for a clear policy outcome to arise directly from scientific data, but in complex, highly ambiguous contexts that involve value conflicts and great uncertainty, such as those that characterize wicked problems, scientific data alone cannot determine the appropriate course of action. As Pielke (2007) explains:

> Science can help us to understand the associations between different choices and their outcomes. . . . Yet, science is rarely a sufficient basis for selecting among alternative courses of action because desired outcomes invariably involve differing conceptions of the sort of world we want in the future. Whether or not avoiding a particular amount of climate change is desirable, or whether or not the risks of nuclear power or GMOs exceed the benefits, are not issues that can be resolved by science alone, but must instead be handled through political processes characterized by bargaining, negotiation and compromise through the exercise of power (139–40).

Coming to agreement about the kind of world we would like in the context of a changing climate requires evaluation of the costs and benefits of different possible responses in the context of different conceptions of values, and consideration of how these responses will impact various regions, people, and ecosystems. Climate vulnerability and adaptation are highly context-specific, requiring consideration of local bio-geography and existing community assets and liabilities, including social networks, demographic composition, socio-economic characteristics, local knowledge and values, and non-climatic pressures (Lorenzoni, Nicholson-Cole, and Whitmarsh 2007; Preston et al. 2011; Wolf 2011). Civic participation and deliberation are key to capturing such knowledge.

Public Participation and Governance

The essential role of public participation in climate governance has long been recognized. Public participation refers to "organized processes adopted by elected officials, government agencies, or other public- or private-sector organizations to engage the public in environmental assessment, planning, decision making, management, monitoring, and evaluation" (National Research Council 2008, 1). The Earth Summit in Rio de Janeiro in 1992, at which the UN Framework Convention on Climate Change was initially negotiated, clearly identified the key role for public participation in addressing climate change: Article 6 outlines the need for "public access to information on climate change and its effects" and "public participation in addressing climate change and its effects and developing adequate responses" (UN 1992). Over the past few decades the importance of public participation has been reaffirmed in all major UN sustainable development resolutions, and in 2010 the Governing Council of the UN Environment Programme adopted "a voluntary set of guidelines for national legislation on access to information, public participation and access to justice in environmental matters" (Jodoin, Duyck, and Lofts 2015, 118). Likewise, public participation has become an essential component of environmental governance processes in many national, regional, and municipal jurisdictions (e.g., City of Edmonton 2005, UN 1992, and National Research Council 2008).

While many recognize the importance of incorporating public participation in environmental management decisions, it is not always clear what this entails. Public participation scholarship highlights two categories of benefits—process and outcome—and two beneficiaries—citizens and government (Irvin and Stansbury 2004). Theoretical justifications for public participation include normative and pragmatic claims about its value: it reflects people's democratic right to participate in decisions that affect their lives, and increases trust in public policy decisions and civil society, thereby creating greater public buy-in, and results in higher-quality decisions because it incorporates a diversity of values and needs (Few, Brown, and Tompkins 2007; Jodoin, Duyck, and Lofts 2015; Reed 2008). As well, some scholars emphasize the importance of increasing the capacity of citizens to effectively engage in participatory decision-making processes (Dietz 2013). In spite of the breadth of these assumed benefits, few attempts have been made to systematically test the validity of these claims (Reed 2008; Dietz 2013).

Alongside calls for increased public engagement in environmental governance has come recognition of the challenges associated with undertaking effective public participation. The difficulties range from debate over different modes of engagement and the extent to which different engagement forums allow for active and meaningful inclusion of the public, to practical and conceptual challenges in securing broadly based citizen representation and ferment over how to appropriately frame issues (Aklin and Urpelainen 2013; Dietz 2013; Few, Brown, and Tompkins 2008; Hulme 2009a; Jasanoff 2010). In the context of climate change, the challenge is to engage citizens and stakeholders effectively in decision-making processes that address complex, ambiguous, multi-scale wicked issues affecting multiple stakeholders and institutions, while focusing the discussion on both "reactive adaptation (responding to an event) and anticipatory adaptation (responding to an expected or likely event)" (Collins and Ison 2009a, 359). Such situational framing needs to move beyond the emphasis on climate change as a global phenomenon, which can suggest an abstract world emptied of social and cultural context where the differences of geography, class, occupation, and gender are largely erased (Jasanoff 2010). While a global view may be an appropriate abstraction for climate modelling, such an apolitical and aspatial framing fails to acknowledge both the variability of risks and resources in different locations and the abilities of different peoples and species to effectively respond to the impacts of climate change (Jasanoff 2010; Klein 2014; Methmann, Rothe, and Stephan 2013). Most people understand the impacts of climate change "in a situated and relational way" (Chilvers et al. 2014, 174) and connect this phenomenon to other concerns in their everyday lives (Leiserowitz 2006; Lorenzoni and Pidgeon 2006). Consequently, a key challenge in engaging citizens in decisions related to climate adaptation is to move them beyond simply perceiving problems and advocating instead for solutions based on spatial and temporal immediacy (Few, Brown, and Tompkins 2007).

Those tasked with designing public participation forums that acknowledge and build upon such diversity often speak of designing dynamic co-evolutionary social learning processes (Collins and Ison 2009a and 2009b; Dietz 2013; Hale 2010; Reed 2008). As used in climate adaptation research and practice, social learning refers to a collective and communicative learning process that moves away from an individualized, educational emphasis. Social learning highlights the need for iterative learning processes involving multiple kinds of knowledge (scientific, community, political); questioning norms, objectives, and policies; and discussing facts, interests, and values. These forms of reflection enable participants to formulate integrated climate adaptation and mitigation strategies (Bos, Brown, and Farrelly 2013; Collins and Ison 2009b; Robinson and Berkes 2011; van der Wal et al. 2014). Successful social learning should result in stakeholders "developing new collective capacities to deal with common problems...to implement conscious and long term adaptive changes in cognitive frameworks of action, and in institutional arrangements, so as to achieve common goals that would otherwise not be achieved individually" (Tàbara et al. 2010, 2).

Public Deliberation as a Form of Public Participation

Public participation is a very broad term: it includes grassroots democracy, civil society mobilization, and forums organized by governments and other organizations to elicit the views of citizens. One can narrow things somewhat by talking about public involvement and public engagement: here, the emphasis is on forums for citizens convened by governments, businesses, and other organizations, including focus groups, town hall meetings, public consultations, design charrettes, and many other mechanisms; groups like the International Association for Public Participation (IAP2) bring together expertise and resources relevant to this wide range of engagement approaches. Deliberative democracy or public deliberation narrows things still further: deliberative approaches centre on diverse participants reasoning together, hearing a diversity of perspectives, articulating underlying values, weighing trade-offs, and coming to a common decision, supported by good information and a clear sense of the influence their voices will have. Most public consultation processes—at least in Alberta—are only weakly deliberative: they typically do not support sustained reasoning between participants, provide only limited information about the issue and the political context for decisions, and elicit existing beliefs and commitments rather than exposing these to challenges and sustained dialogical exploration.

Public deliberation has existed in different forms in many historical and contemporary contexts, from the Athenian city-state to the town hall meetings of early New England: anywhere that ordinary citizens could meet together to exchange perspectives and reasons about political choices in ways that shaped the actions of government. In contemporary mass politics, though, there is a wide gulf between citizens and governments, and fewer and fewer spaces in which citizen deliberation can influence government decisions. And yet intellectuals, scholars, and activists have kept alive the vision of meaningful, thoughtful citizen involvement as a part of political decision making. John Dewey (1927), writing in the United States, emphasized the importance to democracy of inclusive deliberation by citizens as a counterweight to expert and elite influence on government. And since the 1980s, deliberative democracy has become a core theme in political and social theory.

Deliberative democracy as a social and political theory owes much to Jürgen Habermas, a German social theorist and public intellectual. Habermas criticizes the extent to which political life in modern democracies has been given over to elite decision making, and also to the logics of bureaucracies and markets (Habermas 1987). He offers a philosophically careful articulation of the qualities of collective, communicative reasoning needed to reach legitimate democratic decisions (Habermas 1985), and in later work explores how this reasoning can be achieved through an interplay of informal deliberation in the public sphere and formal decision making by elected bodies (Habermas 1996).

Habermas's interest is in theoretically articulating the nature of good deliberation, and in looking at how deliberation can fit within the major structures of a contemporary liberal democracy. It has fallen to other theorists and social scientists to consider the finer-grained institutions and practices through which public deliberation can take place. Key scholars of more pragmatic dimensions of deliberative democracy include James Bohman (1996), Andrea Cornwall (2008), John Dryzek (2010), Archon Fung (2003), Amy Gutmann and Dennis Thompson (1998), and Iris Young (2001). In a series of debates and interventions, these and many other theorists explore normative questions around democratic deliberation: who needs to be included and what inclusion means; how structural dynamics of marginalization and oppression can diminish or prevent the participation and influence of particular social groups and how this can be mitigated; different forms of expression and reasoning and how these fit into deliberative processes; what kinds of decision procedures can express the will of participants; and how deliberative democratic participation can and should shape the participants themselves as well as broader political processes.

From the late 1990s, scholarship on deliberative democracy has taken a more practical turn, studying particular examples of public deliberation such as citizen juries, citizen assemblies, deliberative polls, and consensus conferences (in the language introduced earlier, these are "mini-publics"). This practical turn draws on theoretical articulations of what public deliberation should be like, and uses the tools of social science to explore the actual structure and dynamics of particular exercises, emphasizing their deliberative quality, impacts on participants, and influence on political decisions (e.g., Gastil and Levine 2005; Johnson 2015; Lee 2015; Nabatchi and Leighninger 2015; Rowe and Frewer 2004). This practical turn in scholarship coincides with a growth in professional networks of dialogue and deliberation practitioners, which also involve academics: for example, the International Association for Public Participation, the National Coalition for Dialogue and Deliberation, the Deliberative Democracy Consortium, and the Canadian Community for Dialogue and Deliberation. A further intersection of research and practice can be found at www.participedia.org, which assembles thousands of examples of participatory initiatives, including data about their design and impacts.

The terminology around deliberative and non-deliberative approaches to public participation is tangled and inconsistent. This section has drawn a distinction between deliberative approaches to public involvement and a broader range of tools that engage citizens but without the emphasis on careful collective reasoning across differences to shape political decisions. This is a spectrum rather than a dichotomy. The most common terms used to describe the deliberative end of the spectrum are deliberative democracy, citizen deliberation, public deliberation, and deliberative dialogue; practical exercises that have strongly deliberative qualities travel under names like consensus conferences, citizen juries, citizen panels, and citizen assemblies. A whole host of terms is used to describe forms of public participation that place less emphasis on deliberation as we've defined it here: public consultation, public engagement, public involvement, and public dialogue. These terms are used differently by different experts and lay people, and this slipperiness of terminology is a challenge when it comes to communicating about public deliberation.

Concerns about Public Deliberation

Theoretical and social scientific literatures on public deliberation tend to characterize it quite positively. On the theoretical side, public deliberation is taken to be a crucial source of democratic legitimacy and public trust, as well as an important way to build civic skills and capacity (Bohman 1996; Gutmann and Thompson 1998; Habermas 1996). Social scientific research often draws on this normative view and substantiates the ability of citizens to wrestle with technically complex issues, learn across diversities of perspective, and come to agreement on paths forward. There also is strong evidence that participants enjoy deliberative processes (Abelson and Gauvin 2006; Rosenberg 2007, though for a cautionary perspective on the quality of this social scientific work see Pincock 2012).

Yet there is a range of critical worries about deliberative democracy, expressed by some of its scholarly advocates (e.g., Bickford 1996; Bohman 1996; Williams 1998; Young 2001) and also by scholars who are skeptical of its benefits or more deeply concerned about its deficits. These critical worries relate to power, inclusion, and marginalization, and to dynamics of power and privilege both between participants and with facilitators and experts (Coelho and von Lieres 2010; Cornwall and Gaventa 2001; Cornwall and Coelho 2007; Fischer 2009; Forester 2009; Gaventa and Barrett 2010; Hendriks 2011). One of the most enduring concerns with deliberative democracy is that notwithstanding the normative aspirations of theorists and the good intentions of practitioners, public deliberation can end up reinforcing rather than challenging power relations among participants; between privileged and marginalized social groups; and between participating citizens, experts, and elites (Bickford 1996; Williams 1998; Young 2001).

Another emerging literature looks at the professionalization of public participation, which is increasingly carried out by credentialed experts and large consultancies, and shaped by norms, networks, and trainings that characterize an increasingly networked field. A particularly astringent critique is offered by

Carolyn Lee (2015): she suggests that while professional facilitators and process designers speak consistently of the empowering effects of participation and the need to tailor it to particular contexts, in fact there is an overall sameness to the repertoire of techniques used. Moreover, she argues that participation exercises often provide the illusion of democratic influence, while in fact fitting smoothly with neoliberal state power and managerialism. Genevieve Johnson (2015) also offers a pessimistic view of the influence of public deliberation: looking at four Canadian case studies of well-resourced and organized deliberations, she suggests that they had very limited impact on elite decision making.

A Systems Approach to Public Deliberation

A further scholarly current around public deliberation is deliberative systems theory. Deliberative systems theorists criticize the focus on particular mini-publics, and urge attention instead to how deliberative norms can be met at the level of a whole political system. In some ways this is continuous with Habermas's (1996) interest in how formal and informal institutions of democracy can fit into a functioning democratic whole. Deliberative systems theorists draw more explicitly on scholarship about mini-publics and are more focused on the particular institutional shape of deliberation in a complex democracy. In the words of Mansbridge et al.,

> To understand the larger goal of deliberation, we suggest that it is necessary to go beyond the study of individual institutions and processes to examine their interaction in the system as a whole. We recognize that most democracies are complex entities in which a wide variety of institutions, associations, and sites of contestation accomplish political work—including informal networks, the media, organized advocacy groups, schools, foundations, private and non-profit institutions, legislatures, executive agencies, and the courts. We thus advocate what may be called a systemic approach to deliberative democracy. (2012, 1–2)

Deliberative systems theory enables a more differentiated understanding of mechanisms of deliberative democracy and governance. And it offers a more subtle normative evaluation of deliberative settings. Hayley Stevenson and John Dryzek (2014) provide one of the most comprehensive articulations of how to assess the deliberative quality of a political system, suggesting one needs to look at the health of:

1. The public space (the range of narratives and views expressed in the media, civil society, and among citizens)

2. The empowered space (the range of views interacting in legitimate spaces of collective political decision, e.g. parliaments, courts)

3. The formal and informal transmission of views and narratives between public and empowered spaces

4. The accountability of empowered spaces to public space (through elections, transparency mechanisms, public hearings, etc.)

5. Private spaces (and how well views arising in non-civic spaces are transmitted to public and empowered spaces)

6. Meta-deliberation on the deliberative quality of the system as a whole, and how well it is reflecting the diversity of narratives and discourses in society

7. The decisiveness of the deliberative system in yet broader systems—that is, whether it yields outcomes that affect people's lives (Stevenson and Dryzek 2014, 27–29).

This approach requires an assessment of dynamics and influences that go far beyond the boundaries of any particular mini-public. And the reach of these seven elements of democracy provides a backdrop for worrying about the ambivalent impact of mini-publics.

Simon Burall draws out the potentially negative effects of mini-publics on broader democratic and governance systems. Deliberative mini-publics are usually "conceived of, framed and run by those in authority." As a result, participation "can be used to reinforce authority rather than challenge it, becoming a management tool for securing legitimacy about specific decisions of or institutions themselves" (Burall 2015, 22–23). Moreover, "because much contemporary participation is generated by those in authority, at isolated points throughout the system, the risk is that citizen energy and participation is diffused and prevents the development of the forms of mass participation that were successful in pushing for change in earlier decades" (Burall 2015, 23). Approaching public deliberation as a system, rather than as a series of discrete initiatives, guards against the tendency to expect too much of specific mini-publics. Deficiencies in one part of the system can be made up in other areas (though for a critique of this functional way of assessing deliberativeness, see Owen and Smith 2015).

Alberta Climate Dialogue in the Context of Deliberative Democracy

ABCD was initiated by a group of public deliberation scholars and practitioners who wanted to make a difference to climate responses in Alberta, and also to generate learning that could advance the field of deliberative democracy.

ABCD as a whole thus walked interesting lines when it came to relating its work to the diversity of the field, and especially to critical worries about mini-publics and public deliberation. On the one hand, the project sought to make space for critical inquiry and reflection, especially in three major team workshops; moreover, a number of participants in these workshops were scholars working within these critical literatures (Blue and Medlock 2014; Blue, Medlock, and Einseidel 2013; Gaventa 2006; Kahane et al. 2013; Parkins and Mitchell 2005; Parkins 2006; von Lieres and Kahane 2007), and key practitioners in the project also brought a critical eye to the work. On the other hand, a commitment to supporting climate responses, the exigencies of building and holding deliberations, and passion about public deliberation often shaved the edges off these critical worries, or moved them to the level of design decisions within mini-publics, rather than big questions about the systemic role of mini-publics. The chapters that follow thus provide interesting studies in the relationship of deliberative theory to practice, and of the challenges of sustaining a scholar's version of critical engagement when enmeshed in complex and demanding community-based and practice-based action research.

Structure of the Book

This volume is intended to inform scholars, students, public participation and deliberation practitioners, and public officials interested in democratic deliberation and environmental governance. It is designed as an academic collection that engages with theory and social policy, and simultaneously as a resource for practitioners and decision makers who seek insights and techniques related to public deliberation on wicked issues like climate change. The book is organized into eight chapters that explore the strengths, limitations, and challenges of using deliberative methods as an approach to public engagement and decision making about wicked policy problems. To highlight the situated perspectives of the contributors, each chapter includes a short description of the author(s) and their key roles in ABCD. As well, each chapter connects theories and practices of both public deliberation and climate change politics, demonstrating to readers key issues that arise in considering and designing deliberative initiatives.

We built a companion website that provides multiple ways of engaging with this book. While some readers interested in the challenges in using deliberative dialogue to engage citizens about wicked issues, and in reflecting on big picture questions and how they influence particular choices, may want to read the entire book, those wanting a more accessible way into the discussion can use the website, which links to each individual chapter and also offers a quick overview of the entire book, chapter "take-aways," and beginner and advanced resource lists on each topic.

In Chapter 1, Lorelei Hanson provides a profile of the four citizen deliberation projects members of ABCD actively participated in from 2012 to 2014. Hanson highlights the unique features of each public deliberation and the successes and challenges in realizing key social learning outcomes to allow for collective agreement and action. The voices of citizens and the volunteer small group facilitators and note takers are emphasized in this chapter, thus integrating their experiences and views into the critical evaluation of each of the four ABCD deliberations.

Chapter 2 expands on the discussion of public deliberation provided in the introduction. David Kahane and Gwendolyn Blue outline the promise of, and key debates within, deliberative democracy theory and practice, and trace how these were reflected in ABCD's work. They explore how climate change poses particular challenges to deliberative approaches, including those around framing, representation, and the politics of knowledge.

Chapter 3 describes the political and economic contexts within which ABCD operated. As Geoff Salomons and John Parkins demonstrate, an understanding of history and context is key to successfully designing and facilitating effective and meaningful public deliberation. They describe how upper levels of government were reluctant to undertake citizen deliberation on climate change because it did not align with ideological positions and political goals, and why there was more interest and uptake at the municipal level. The authors also illustrate how municipal governments are most immediately impacted by climate change and require large-scale buy-in from citizens to move forward on complex social policy issues; they therefore can be willing to undertake deliberative citizen involvement even when outcomes are unsure and may threaten powerful interests.

Chapter 4 explores how climate change is framed and presented and how this influences dialogue and action. While deliberation typically focuses on individual ideas, interests, and values, Gwendolyn Blue demonstrates how these are shaped by language and discourse. Blue highlights two prominent global discourses of climate change—mitigation and adaption—and how these climate discourses influence public deliberation, including how the problem is conceptualized and the solutions and actions open for consideration. She argues that organizers of public deliberation on climate change should identify and integrate the range of issue frames and options that interested and affected parties consider viable.

Chapter 5 examines how participants are recruited for deliberative exercises. Shelley Boulianne highlights key theoretical debates and methodological issues associated with recruiting for representativeness and inclusivity. She analyzes recruitment for the four public deliberations ABCD participated in, revealing challenges in ensuring that a truly representative group of citizens is convened, and discussing trade-offs in representativeness when inclusion of minority voices is a key objective. She concludes that a focus on climate change complicates recruitment for public deliberation. Too often, citizens engage in self-selection when wicked and complex scientific issues are at stake, resulting in biases that impact the policy recommendations that arise out of deliberative exercises.

Chapter 6 discusses the essential role of collaboration in public deliberation projects, particularly those focused on social learning to effectively address wicked issues. Collaborators bring divergent knowledge, norms, ways of navigating political bureaucracies, and communication styles, as well as conflicting allegiances and identifications. David Kahane and Lorelei Hanson explore challenges, tensions, strengths, and opportunities that arose in the four ABCD-linked deliberation projects. They suggest that collaboration was most successful when parties were strongly invested in outcomes, when communication was open, and where there was sufficient time to develop mutual trust.

Chapter 7 presents the perspective of experienced deliberation practitioners Mary Pat MacKinnon, Jacquie Dale, and Susanna Haas Lyons, who were centrally involved in designing and facilitating three of ABCD's deliberations. They explore challenges associated with framing a topic for deliberation, particularly a topic as complex as climate change. They describe techniques for practically addressing the deliberation context, seizing opportunities for impact, and addressing partners' expectations. The chapter highlights the role of values in public deliberation, how to manage differences in topical knowledge and ways of knowing/learning, and how to work with dynamics of ownership and power within groups of participants. The chapter authors provide a useful overview of both the flexibility and rewards of public deliberation, and challenges in designing and facilitating processes that produce authentic and useful results with impact for both citizens and decision makers.

Chapter 8 looks more broadly at strengths and limitations of deliberative democracy in addressing complex systemic problems. David Kahane outlines eight stories of social change told within the deliberative democracy community in order to critically evaluate ABCD's impact. He argues that neither the field of deliberative democracy nor ABCD has sufficiently focused on a whole systems approach, and considers why this is the case. Through a review of the key insights from the fields of systems thinking, user-centred design, and systemic design, he demonstrates benefits of methodically and consistently organizing complex deliberations around questions of systems change, in terms of the orientation of the overall project, the development of particular partnerships, and the design of citizen deliberations. In this way, he provides a useful set of considerations for those embarking on public deliberation projects to more effectively address wicked issues like climate change.

The conclusion provides a short overview of the key themes and observations highlighted in each of the preceding chapters. Tom Prugh and Matt Leighninger discuss the role public deliberation could play in an increasingly activist urban and community-centred society that is grappling with profound shifts in climate, the economy, and energy systems. While recognizing public deliberation is not without its limitations and problems, the authors argue that it offers a method for citizens to come to grips with wicked issues like climate change that are both universal and particular. Deliberation combined with local action provides a potent combination for sustained engagement through which citizens can anticipate and cope with the coming environmental challenges wrought by climate change, and thereby provides a means for strengthening community capacity at many levels.

ABCD as a project came together to advance the field of public deliberation, and to explore how public deliberation can advance public and state responses to the challenge of climate change. We hope that this volume communicates our learning in ways that help others to engage the public on the many grave and wicked problems facing our societies.

References

Abelson, Julia, and Francois-Pierre Gauvin. 2006. *Assessing the Impacts of Public Participation: Concepts, Evidence, and Policy Implications*. Ottawa: Canadian Policy Research Networks.

Adkin, Laurie E., Lorelei L. Hanson, David Kahane, John R. Parkins, and Steve Patten. 2016. "Can Public Engagement Democratize Environmental Policy-Making in a Resource-Dependent State? Comparative Case Studies from Alberta, Canada." *Environmental Politics*. doi:10.1080/09644016.2016.1244967.

Aklin, Michaël, and Johannes Urpelainen. 2013. "Debating Clean Energy: Frames, Counter Frames, and Audiences." *Global Environmental Change* 23: 1225–32.

APSC (Australian Public Service Commission). 2007. *Tackling Wicked Problems: A Public Policy Perspective*. http://www.apsc.gov.au/publications-and-media/archive/publications-archive/tackling-wicked-problems.

Beck, Ulrich. 2010. "Climate for Change, or How to Create a Green Modernity?" *Theory, Culture and Society* 27(2–3): 254–66.

Bickford, Susan. 1996. *The Dissonance of Democracy*. Ithaca, NY: Cornell University Press.

Blue, Gwendolyn, and Jennifer Medlock. 2014. "Public Engagement with Climate Change as Scientific Citizenship: A Case Study of World Wide Views on Global Warming." *Science as Culture* 23(4): 560–79.

Blue, Gwendolyn, Jennifer Medlock, and Edna Einsiedel. 2013. "Representativeness and the Politics of Inclusion: Insights from WorldWideViews Canada." In *Citizen Participation in Global Environmental Governance*, edited by Richard Worthington, Mikko Rask, and Minna Lammi, 139–52. London: Earthscan.

Bohman, James. 1996. *Public Deliberation: Pluralism, Complexity and Democracy*. Cambridge: MIT Press.

Bos, Annette, Rebecca R. Brown, and Megan A. Farrelly. 2013. "A Design Framework for Creating Social Learning Situations" *Global Environmental Change* 23(2): 398–412.

Brulle, Robert J., and Riley E. Dunlap. 2015. "Sociology and Global Climate Change: Introduction." *Climate Change and Society – Sociological Perspectives*, edited by Riley E. Dunlap and Robert J. Brulle, 1–31. New York: Oxford University Press.

Burall, Simon. 2015. *Room for a View: Democracy as a Deliberative System*. London: Involve. http://www.involve.org.uk/wp-content/uploads/2015/10/Room-for-a-View.pdf.

Chilvers, Jason, Irene Lorenzoni, Geraldine Terry, Paul Buckley, John K. Pinnegar, and Stefan Gelcich. 2014. "Public Engagement with Marine Climate Change Issues: (Re)framings, Understandings and Responses." *Global Environmental Change* 29: 165–79.

City of Edmonton. 2005. City of Edmonton Public Involvement, Policy C513. November 18. http://www.edmonton.ca/city_government/documents/PDF/city_procedure.pdf#search=Public%20participation.

Coelho, Vera Schattan, and Bettina von Lieres. 2010. *Mobilizing for Democracy: Citizen Action and the Politics of Public Participation*. London: Zed Books.

Collins, Kevin, and Ray Ison. 2009a. "Jumping off Arnstein's Ladder: Social Learning as a New Policy Paradigm for Climate Change Adaptation." *Environmental Policy and Governance* 19: 358–573.

———. 2009b. "Living with Environmental Change: Adaptation as Social Learning." *Environmental Policy and Governance* 19: 351-357.

Cornwall, Andrea. 2008. *Democratising Engagement: What the UK Can Learn from International Experience*. London: Demos.

Cornwall, Andrea, and John Gaventa. 2001. *From Users and Choosers to Makers and Shapers: Repositioning Participation in Social Policy*. Brighton: Institute of Development Studies.

Cornwall, Andrea, and Vera Schattan Coelho, eds. 2007. *Spaces for Change? The Politics of Citizen Participation in New Democratic Arenas*. London: Zed Books.

Dewey, John. 1927. *The Public and Its Problems*. Athens: Ohio University Press.

Dietz, Thomas. 2013. "Bringing Values and Deliberation to Science Communication." *Proceedings of the National Academy of Sciences* 110(3): 14081–87.

Dryzek, John. 2010. "Mini-publics and their Macro Consequences." In *Foundations and Frontiers of Deliberative Governance*, edited by John Dryzek, 156–77. Oxford: Oxford University Press.

Few, Roger, Katrina Brown, and Emma L. Tompkins. 2007. "Public Participation and Climate Change Adaptation." *Climate Policy* 7(1): 46–59. doi:10.1080/14693062.2007.9685637.

Fischer, Frank. 2009. *Democracy and Expertise: Reorienting Policy Inquiry*. Oxford: Oxford University Press.

Forester, John. 2009. *Dealing with Differences: Dramas in Mediating Public Disputes*. Oxford: Oxford University Press.

Fung, Archon. 2003. "Survey Article: Recipes for Public Spheres: Eight Institutional Design Choices and Their Consequences." *Journal of Political Philosophy* 11(3): 338–67.

Gastil, John, and Peter Levine, eds. 2005. *The Deliberative Democracy Handbook*. San Francisco: Jossey-Bass.

Gaventa, John. 2006. "Finding Spaces for Change: A Power Analysis." *Institute for Development Studies Bulletin* 37(6): 23–33.

Gaventa, John, and Greg Barrett. 2010. "So What Difference Does It Make? Mapping the Outcomes of Citizens Engagement." *Working Paper* No. 347. Brighton, UK: Institute of Development Studies.

Gutmann, Amy, and Dennis Thompson. 1998. *Democracy and Disagreement*. Cambridge: Harvard University Press.

Habermas, Jürgen. 1985. *The Theory of Communicative Action*. Vol. 1: *Reason and Rationalization of Society*. Translated by Thomas McCarthy. Boston: Beacon Press.

———. 1987. *The Theory of Communicative Action*. Vol. 2: *Lifeworld and System: A Critique of Functionalist Reason*. Translated by Thomas McCarthy. Boston: Beacon Press.

———. 1996. *Between Facts and Norms*. Translated by William Rehg. Oxford: Polity Press.

Hale, Stephen. 2010. "The New Politics of Climate Change: Why We are Failing and How We Will Succeed." *Environmental Politics* 19(2): 255–75.

Harvey, Fiona. 2015. "Paris Climate Change Deal Too Weak to Help Poor, Critics Warn." *The Guardian*, December 14. http://www.theguardian.com/environment/2015/dec/14/paris-climate-change-deal-cop21-oxfam-actionaid.

Hendriks, Carolyn. 2011. *The Politics of Deliberation: Citizen Engagement and Interest Advocacy*. London: Palgrave Macmillan.

Hernández, Ariel Macaspac. 2014. "Complexities in Global Climate Talks: Stumbling Blocks to Decision Making." In *Strategic Facilitation of Complex Decision-Making*, edited by Ariel Macaspac Hernández, 81–111. Switzerland: Springer International.

Hulme, Mike. 2009a. *Why We Disagree About Climate Change: Understanding Controversy, Inaction and Opportunity*. New York: Cambridge University Press.

———. 2009b. "The Science and Politics of Climate Change." *Wall Street Journal*, December 2. http://www.wsj.com/articles/SB10001424052748704107104574571613215771336.

IPCC (Intergovernmental Panel on Climate Change). 2014. *Climate Change 2014 Synthesis Report Summary for Policymakers*. http://www.ipcc.ch/pdf/assessment-report/ar5/syr/AR5_SYR_FINAL_SPM.pdf.

———. 2015. *Climate Change 2015 Synthesis Report*. http://ar5-syr.ipcc.ch/ipcc/ipcc/resources/pdf/IPCC_SynthesisReport.pdf.

Irvin, Renée A., and John Stansbury. 2004. "Citizen Participation in Decision Making: Is It Worth the Effort?" *Public Administration Review* 64(1): 55–65.

Jasanoff, Sheila. 2010. "A New Climate for Society." *Theory, Culture and Society* 27(2–3): 233–53.

Jodoin, Sébastien, Sébastien Duyck, and Katherine Lofts. 2015. "Public Participation and Climate Governance: An Introduction." *Review of European Community and International Environmental Law* 24(2): 117–22. doi:10.111/reel.12126.

Johnson, Genevieve Fuji. 2015. *Democratic Illusion: Deliberative Democracy in Canadian Public Policy*. Toronto: University of Toronto Press.

Kahane, David, Kristjana Loptson, Jade Herriman, and Max Hardy. 2013. "Stakeholder and Citizen Roles in Public Deliberation." *Journal of Public Deliberation* 9(2): 2.

Kent, Gordon. 2015. "Climate-Change Deniers Should Face Probe: U of A Professor." *Edmonton Journal*, December 7, A4.

Klein, Naomi. 2014. *This Changes Everything: Capitalism versus the Climate*. New York: Simon and Schuster.

Kolko, Jon., 2012. "Wicked Problems: Problems Worth Solving." *Stanford Social Innovation Review*, March 6. http://ssir.org/articles/entry/wicked_problems_ problems_worth_solving.

Land Institute. 2017. "About Us." Accessed March 17. https://landinstitute.org/about-us.

Lawton, Geoff. 2017. "About Geoff Lawton." Accessed March 17. http://geofflawton. com/about/.

Lee, Carolyn. 2015. *Do-It-Yourself Democracy: The Rise of the Public Engagement Industry*. Oxford: Oxford University Press.

Leiserowitz, Anthony. 2006. "Climate Change Risk Perception and Policy Preferences: The Role of Affect, Imagery, and Values." *Climate Change* 77: 45–72.

Levin, Kelly, Benjamin Cashore, Steven Berstein, and Graeme Auld. 2012. "Overcoming the Tragedy of Super Wicked Problems: Constraining Our Future Selves to Ameliorate Global Climate Change." *Policy Sciences* 45(2): 123–52.

Lorenzoni, Irene, Sophie Nicholson-Cole, and Lorraine Whitmarsh. 2007. "Barriers Perceived to Engaging with Climate Change among the UK Public and their Policy Implications." *Global Environmental Change* 17(3–4): 445–59.

Lorenzoni, Irene, and Nick F. Pidgeon. 2006. "Public Views on Climate Change: European and USA Perspectives." *Climatic Change* 77(1): 73–95. doi:10.1007/s10584-006-9072-z.

Lukacs, Martin. 2015. "Claim No Easy Victories. Paris Was a Failure, But a Climate Justice Movement Is Rising." *The Guardian*, December 15. http://www.theguardian. com/environment/true-north/2015/dec/15/claim-no-easy-victories-paris-was-a-failure-but-a-climate-justice-movement-is-rising.

Mandel, Charles. 2015. "Ecojustice Files Complaint with Competition Bureau Against Climate Denial Groups." *National Observer*, December 3. http://www. nationalobserver.com/2015/12/03/news/breaking-ecojustice-files-complaint-competition-bureau-against-climate-denial-groups.

Mansbridge, Jane, James Bohman, Simone Chambers, Thomas Christiano, Archon Fung, John Parkinson, Dennis F. Thompson, and Mark E. Warren. 2012. "A Systemic Approach to Deliberative Democracy." In *Deliberative Systems: Deliberative Democracy at the Large Scale*, edited by John Parkinson and Jane Mansbridge, 1–26. Cambridge: Cambridge University Press.

Methmann, Chris, Delf Rothe, and Benjamin Stephan. 2013. "Introduction: How and Why to Deconstruct the Greenhouse." In *Interpretive Approaches to Global Climate Governance: (De)constructing the Greenhouse*, edited by Chris Methmann, Delf Rothe and Benjamin Stephan, 1–22. New York: Routledge.

Monbiot, George. 2015. "Grand Promises of Paris Climate Deal Undermined by Squalid Retrenchments." *The Guardian*, December 12. http://www.theguardian.com/environment/georgemonbiot/2015/dec/12/paris-climate-deal-governments-fossil-fuels.

Moser, Susanne C. 2010. "Communicating Climate Change: History, Challenges, Process and Future Direction." *Wiley Interdisciplinary Review: Climate Change* 1(1): 31–53. doi:10.1002/wcc.011.

Nabatchi, Tina, and Matt Leighninger. 2015. *Public Participation for the 21st Century*. San Francisco: Jossey-Bass.

National Research Council. 2008. *Public Participation in Environmental Assessment and Decision Making*. Panel on Public Participation in Environmental Assessment and Decision Making, edited by Thomas Dietz and Paul C. Stern. Committee on the Human Dimensions of Global Change. Division of Behavioral and Social Sciences and Education. Washington, DC: National Academies Press. http://www.nap.edu/catalog/12434.html.

Owen, David and Graham Smith. 2015. "Survey Article: Deliberation, Democracy, and the Systemic Turn." *Journal of Political Philosophy* 23(2): 213–234.

Parkins, John. 2006. "De-centering Environmental Governance: A Short History and Analysis of Democratic Processes in the Forest Sector of Alberta, Canada." *Policy Sciences* 39(2): 183–202.

Parkins, John, and Ross E. Mitchell. 2005. "Public Participation as Public Debate: A Deliberative Turn in Natural Resource Management." *Society and Natural Resources* 18(6): 529–40.

Pielke, Roger A., Jr. 2007. *The Honest Broker: Making Sense of Science in Policy and Politics*. New York: Cambridge University Press.

Pincock, Heather. 2012. "Does Deliberation Make Better Citizens?" In *Democracy in Motion: Evaluating the Practice and Impact of Deliberative Citizen Engagement*, edited by Tina Nabatchi, John Gastil, G. Michael Weiksner, and Matt Leigninger, 135–62. Oxford: Oxford University Press.

Pinker, Steven. 2015. "Now for the Good News: Things Really Are Getting Better." *The Guardian*, September 11. http://www.theguardian.com/commentisfree/2015/sep/11/news-isis-syria-headlines-violence-steven-pinker.

Preston, Benjamin L., Richard M. Westaway, and Emma J. Yuen. 2011. "Climate Adaptation Planning in Practice: An Evaluation of Adaptation Plans from Three Developed Nations." *Mitigation and Adaptation Strategies for Global Change* 16(4): 407–38.

Reed, Mark S. 2008. "Stakeholder Participation for Environmental Management: A Literature Review." *Biological Conservation* 141: 2417–31. doi:10:1016/j.biocon.2008.07.014.

République Française. 2015. United Nations Conference on Climate Change. http://www.cop21.gouv.fr/en/more-details-about-the-agreement/.

Riedy, Chris. 2013. "Climate Change Is a Super Wicked Problem." *Planetcentric*, May 29. http://chrisriedy.me/2013/05/29/climate-change-is-a-super-wicked-problem/.

Rittel, Horst W.J., and Melvin M. Webber. 1973. "Dilemmas in a General Theory of Planning." *Policy Sciences* 4: 155–69.

Robinson, Lance W., and Fikret Berkes. 2011. "Multi-level Participation for Building Adaptive Capacity: Formal Agency-Community Interactions in Northern Kenya." *Global Environmental Change* 21(4): 1185–94.

Rosenberg, Shawn W., ed. 2007. *Democracy, Deliberation and Participation: Can the People Decide?* London: Palgrave Macmillan.

Rowe, Gene, and Lynn J. Frewer. 2004. "Evaluating Public-Participation Exercises: A Research Agenda." *Science, Technology and Human Values* 29(4): 512–56.

Stevenson, Hayley, and John Dryzek. 2014. *Democratizing Global Climate Governance*. Cambridge: Cambridge University Press.

Szersynski, Bronislaw, and John Urry. 2010. "Changing Climates: Introduction." *Theory, Culture and Society* 27(2–3): 1–8.

Tàbara, J. David, Xingang Dai, Gensuo Jia, Darryn McEvoy, Henry Neufeldt, Anna Serra, Saskia Werners, and Jennifer J. West. 2010. "The Climate Learning Ladder. A Pragmatic Procedure to Support Climate Adaptation." *Environmental Policy and Governance* 20: 1–11.

UN (United Nations). 1992. Rio Declaration on Environment and Development.http://www.unep.org/documents.multilingual/default.asp?documentid=78&articleid=1163.

———. 2015. Framework Convention on Climate Change. December 12. http://unfccc.int/documentation/documents/advanced_search/items/6911.php?priref=600008831.

van der Wal, Merel, Joop De Kraker, Astrid Offermans, Carolien Kroeze, Paul A. Kirschner, and Martin van Ittersum. 2014. "Measuring Social Learning in Participatory Approaches to Natural Resource Management." *Environmental Policy and Governance* 24: 1–15.

von Lieres, Bettina, and David Kahane. 2007. "Inclusion and Representation in Democratic Deliberations: Lessons from Canada's Romanow Commission" in *Spaces for Change?*, edited by Cornwall and Schattan Coelho, 131–51. London: Zed Books.

Williams, Melissa. 1998. *Voice, Trust, and Memory: Marginalized Groups and the Failings of Liberal Representation*. Princeton, NJ: Princeton University Press.

Wolf, Johanna. 2011. "Climate Change Adaptation as a Social Process." In *Climate Change Adaptation in Developed Nations: From Theory to Practice*, edited by James Ford and Lea Berrang Ford, 21–32. London: Springer. doi 10.1007/978-94-007-0567-82.

Young, Iris Marion. 2001. *Inclusion and Democracy*. Oxford: Oxford University Press.

Profiles of Four Citizen Deliberations

Lorelei L. Hanson

Effective engagement with a broad range of citizens is at the heart of public deliberation focused on realizing principles of inclusion, equality, information, and reason. These principles are integral to ensuring democratic empowerment that provides citizens "the capacities, capabilities and opportunities" to directly "influence public policies" (Johnson 2009, 680). Yet deliberations focused on wicked issues present an additional layer of complexity. Intractable problems that involve competing values and tensions—where time is not costless and those most responsible for the problem have the least immediate incentive to do something about it—challenge existing public policy engagement processes at many levels (Lazarus 2009; Levin et al. 2012; see introduction). Recognition of such complexity has likewise emerged in calls for a new way of approaching how we manage our interactions with natural ecosystems. Kay and Schneider (1994, 32) explain:

> Scientific judgments about right and wrong seemed possible when we
> viewed the world as a set of billiard balls . . . . Unfortunately, this worldview
> with its approach to governance and law does not recognize, and will not
> help us deal with, the realities of complex systems. And here we have the
> crux of the issue. If we are truly to use an ecosystem approach, and we must
> if we are to have sustainability, it means changing in a fundamental way
> how we govern our decision-making processes and institutions, and how we
> approach the business of environmental science and management.

Within environmental management there is increasing recognition of the need for anticipatory, adaptive, and community-based approaches to address the complexity and dynamic nature of socio-natural systems (Diduck et al. 2012; Reed 2008; Tompkins and Adger 2004; Waltner-Toews et al. 2003; Wilner et al. 2012). Adaptive management is "a systematic process for improving management policies by learning from the outcomes of management strategies that have already been implemented" (Pahl-Wostl et al. 2007). Addressing super wicked issues like climate change from such a perspective directs one to consider how to build institutional designs that are not path dependent, but rather flexible, inclusive, and iterative procedures that allow for the development of continuously emergent management approaches. Key to this process is social learning that occurs at both individual and collective levels (Diduck et al. 2012) and allows participants to "monitor the outcome of their decisions and adapt them accordingly" (Reed 2008, 2422).

Social learning theory is informed by a number of disciplines and understandings of how learning occurs (Bandura 1977; Baron and Kerr 2003; Lave and Wenger 1991). Social learning emphasizes that "cognition is not solely an internalized, psychological process, but is essentially context-dependent and interactive" (Muro and Jeffrey 2008, 328). Key to many conceptualizations of social learning is a focus on observing and modelling behaviours, attitudes and emotional reactions, reciprocity and feedback, and social participation. There is no agreed-upon definition of social learning, but within natural resource management scholarship there are some generally agreed-upon outcomes generated from this form of collective and communicative learning:

- New factual knowledge
- Technical and social skills
- Change of cognition and attitudes
- Development of trust and formation of relationships.

Ultimately, these outcomes should result in collective agreement and action; "social learning is not only seen as a prerequisite for individual behavioural change but also for collective action" (Muro and Jeffrey 2008, 332). Given the focus on building collective cognition and action, social learning theory is particularly suited to citizen deliberation.

This chapter outlines four citizen deliberation projects members of Alberta Climate Dialogue (ABCD) actively participated in from 2012 to 2014 in chronological order: City of Edmonton City-Wide Food and Urban Agriculture Citizen

Panel; Citizens' Panel on Edmonton's Energy and Climate Challenges; Energy Efficiency Choices; and Water in a Changing Climate.

I highlight the unique features of each public deliberation and the successes and challenges in realizing some of the above social learning outcomes, particularly from the vantage point of citizens and the volunteer table hosts and note takers. As one of the core members of ABCD, and a researcher associated with three of the deliberative exercises, I have drawn from my observational notes, as well as documents produced for each deliberation (planning materials, participant handouts, agendas, and final reports), and research data (surveys, and semi-structured and focus group interview transcripts)[1] to outline the structure and activities associated with each deliberation, and the outcomes achieved. The deliberative profiles also include some discussion of a range of factors that were considered in planning the deliberations, and thereby foreground the issues that lie at the heart of more detailed analysis in the remaining chapters of this book.

Edmonton's City-Wide Food and Urban Agriculture Citizen Panel

The Deliberation Design and Unique Features

The first deliberation members of ABCD participated in was Edmonton's City-Wide Food and Urban Agriculture Citizen Panels (Food and Ag Panel). The Food and Ag Panel was convened by the Centre for Public Involvement (CPI), a small not-for-profit organization jointly funded by the University of Alberta and City of Edmonton to provide leadership on public participation. Although initial meetings focused on ABCD collaborating with CPI on this project, in the end ABCD played a very minor role because of the constrained project time frame and the politically sensitive context within which this deliberation emerged. Members of ABCD provided some initial recommendations on framing and ongoing research support: two ABCD researchers helped formulate questions for the citizen surveys, and another two observed the deliberation and undertook semi-structured in-depth interviews with five of the citizen panelists. In spite of ABCD's limited role in the design of the deliberation, the hope was that ABCD could learn from CPI's experiences (see illustration on pp. 36–37).

EDMONTON'S FOOD and URBAN AGRICULTURE

CITIZENS' PANEL

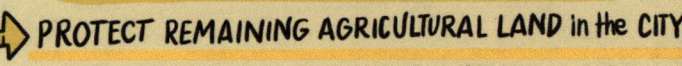

CONTEXT
HIGHLY POLITICIZED

CIVIL SOCIETY ACTIVISM (SINCE NOV 2008)
- MOBILIZES PUBLIC TO PRESSURE CITY COUNCIL TO:

➡ PROTECT REMAINING AGRICULTURAL LAND in the CITY

➡ DEVELOP A FOOD & AGRICULTURE STRATEGY for EDMONTON

1 MAY 2011

MAYOR RESPONDS:
APPOINTS 15 MEMBER CITY-WIDE FOOD and AGRICULTURE STRATEGY ADVISORY COMMITTEE

2 APRIL - JUNE 2012

SUSTAINABLE DEVELOPMENT (CITY OF EDM) and CENTRE FOR PUBLIC INVOLVEMENT (CPI) DESIGN AND CARRY OUT CITIZENS' DELIBERATION PANEL TO FEED INTO THE PROCESS TO DEVELOP STRATEGY

57 CITIZENS DELIBERATE

6 SESSIONS
2 FULL DAY and 4 HALF DAY WORKING GROUPS

3

THE THEME of EACH WEEK

1) INTRO: WELCOME and DISCUSS VALUES
2) HEALTHY FOOD and FOOD SECURITY
3) VIBRANT and DIVERSE ECONOMY
4) ENERGY REDUCTION and HEALTH ECOSYSTEMS
5) ATTRACTIVE and VIBRANT PLACES
6) FINALIZE and PRIORITIZE RECOMMENDATIONS

JULY 2012

4 CITIZENS' PANEL REPORT WRITTEN

TOP Recommend-ations

Protect remaining agricultural land in city

Provide more space for urban agriculture

Develop and support more market space and food hubs

Expand and support urban agriculture through education and policy

ABCD provided limited input on research plan: observed deliberation and interviewed key organizer and 5 citizens

Report

REPORT

REPORT SENT TO CITY WIDE FOOD and AGRICULTURE STRATEGY ADVISORY COMMITTEE

5 **OCTOBER 2012**

SUSTAINABLE DEVELOPMENT and HB Lanarc WRITE A DRAFT STRATEGY

CITY-WIDE FOOD and AGRICULTURE STRATEGY ADVISORY COMMITTEE APPROVES STRATEGY WHICH IS THEN TITLED "*fresh*"

"*fresh*" SUBMITTED TO CITY COUNCIL, CITIZENS APPEAR AT COUNCIL HEARINGS *including* MEMBERS OF CITIZENS' PANEL

6 **NOVEMBER 2012**

CITY COUNCIL APPROVES *fresh* AND THE ACCOMPANYING DECISIONS:

○ *most of* THE REMAINING AGRICULTURAL LANDS in Edmonton IDENTIFIED for RESIDENTIAL & INDUSTRIAL DEVELOPMENT

• ESTABLISH a FOOD and URBAN AGRICULTURE COUNCIL AND COMMITTED $150,000 annually & 1 FULL TIME STAFF POSITION

*CPI designed, organized and hosted the deliberation with some input from Sustainable Development

The Food and Ag Panel was one of eight public forums convened in the development of *fresh*, Edmonton's food and urban agriculture strategy (City of Edmonton 2012); these forums engaged more than 3,300 citizens and stakeholders (Beckie, Hanson, and Schrader 2013). The wide range of public involvement opportunities associated with *fresh* reflected a formal commitment by the City of Edmonton to engage citizens on issues that affect them (City of Edmonton 2006) but also the "highly politicized context" (Cavanagh 2015) surrounding the development of a food and urban agricultural strategy.

The controversy surrounding the development of *fresh* largely concerned the rezoning of land in Edmonton's northeast from agricultural to residential and commercial. In 1982 Edmonton annexed approximately 8,000 hectares of land bordering the northeast, southeast, and southwest parts of the city, and designated these as future Urban Growth Areas (UGAS). Much of this land was zoned agricultural, and largely remained so for over thirty years, making Edmonton one of the few urban municipalities in Canada with a large amount of prime agricultural land within its boundaries (HB Lanarc Consultants 2012). Of the three UGAS, the northeast lands, received the most public attention, as many considered this area to have particularly favourable growing conditions, with a combination of high-quality soils, unique microclimate, and the potential for irrigation due to its proximity to the North Saskatchewan River (Nutter, Hubbard, and Nutter 2011).

The extensive public engagement process that accompanied the development of *fresh* involved a number of major players who had to complete their work within a very short time span. The City of Edmonton sponsored the process and appointed the planning department to oversee the development of *fresh* over the course of a year and a half. HB Lanarc Consultants, a Vancouver-based planning and design firm that works with local and regional governments and developers on sustainable community and regional planning, was hired to assist with the engagement process. The mayor selected fifteen local stakeholders to serve on an advisory committee that, with the support of HB Lanarc and several staff from the City of Edmonton's planning department, was tasked with developing a draft food and urban agriculture strategy.

The citizen deliberations included two full days (the initial and final sessions) and four half-day Saturday sessions that fifty-eight citizens attended. In the first session the panelists identified seven values to guide the process: environmental sustainability; safe, quality food production with ethical treatment of animals; accessible education on food and agriculture for all citizens; food justice with attention to equity, self-sufficiency, transparency, and accessibility; building community; commitment to inclusivity and cultural diversity; and protecting local production. They used these values to frame their discussion of strategies for four goals: strengthening the local economy; protecting the environment; minimizing waste; and creating vibrant, attractive places. At the end of each deliberation day the strategies identified by citizens were submitted to a master writer, who compiled them into a weekly report for the citizens to review (Centre for Public Involvement 2012a). During the final session, with the assistance of a professional facilitator, the citizens reviewed and voted on all the strategies developed for each goal area. A report on the citizen panel process and recommendations, titled *City-Wide Food and Urban Agriculture Strategy: Report on Citizen Panel Process and Recommendations*, was written by CPI staff in consultation with Edmonton's planning department and the mayor's office. The Food and Ag Panel's top two priority recommendations included:

- Municipal leaders "create and/or amend municipal policy tools . . . to prohibit future development on good fertile agricultural land, particularly the northeast farmland"

- "Maximize spaces and places within the City of Edmonton for urban growing and food production." (Centre for Public Involvement 2012a)

A range of research was undertaken throughout the deliberation. During five of the six weeks, citizen panelists were asked to complete paper surveys that tracked their opinions and learning on urban food and agriculture, climate change, broader questions of democratic citizenship and participation, and their knowledge of the Food and Ag Panel process and outcomes. The research design also included a control group of randomly selected Edmontonians who were asked to complete all five of the citizen panel surveys to identify changes in perspectives seen only in the citizen panelists over time. As well, semi-structured interviews were conducted with the lead facilitator after the deliberation and with fifteen individuals all involved in the development of *fresh* including five Food and Ag panelists.

The Food and Ag Panel's report was submitted to City Council and the Advisory Committee, but it had little influence on policy. Given the short time frame, the report was never formally discussed by either City Council or the Advisory Committee (Food and Ag Interview, KI 6 and KI 7). Later, in October 2012, when City Council's executive committee convened a non-statutory public hearing to review the draft food and urban agricultural strategy, of the sixty-three individuals who spoke, no one mentioned the Food and Ag Panel.

The *fresh* strategy received approval in November 2012, thirteen months after the stakeholder advisory committee was appointed. Edmonton's City Council directed City Administration to prepare an implementation plan and budget, and promised continued funding of $150,000 for one full-time position to support a food council (Hanson and Schrader 2014). In February 2013 public hearings were convened to discuss the area structure plan for Edmonton's northeast agricultural lands, which, like the non-statutory meetings for *fresh*, extended over two days due to extensive public interest. In the end, most of the northeast region was rezoned for residential and commercial development to support an adjacent energy and technology park approved in 2010 (Hanson and Schrader 2014).

Social Learning Outcomes

The Food and Ag Panel was designed to feed into the strategy development process by providing a more representative public view of the issues. The goal of the Food and Ag Panel was to "have citizens discuss, learn about, and recommend to City Council strategies about production, distribution and consumption of food" (Centre for Public Involvement 2012b). CPI designed and delivered the deliberation and undertook some research on the process and its impact, but they did not have complete control. Framing of some of the deliberation topics was determined in advance by the planning department in consultation with the mayor, and a couple of times the mayor requested meetings to review the process design. Consequently, as the lead facilitator from CPI explained, they weren't "able to even . . . within the process design of the panel, fully take up the issue of land and land use" (Cavanagh 2015), even though this was clearly an issue of great interest to many citizens.

In spite of the constraints, most citizens indicated overall satisfaction with the deliberation. For example, given the limited time frame, CPI had difficulty administering the recruitment of participants (Torres Scott 2012), and a number of local food activists who were very knowledgeable about urban agriculture were able to take advantage of this predicament and register as panelists. Nonetheless, citizens didn't feel that the process was hijacked by food activists but instead spoke about the diversity of views expressed and how inclusive the process was, as illustrated by these citizen panelists' comments:

The citizen's panel was really cool because it wasn't just all people like me. It was a whole bunch of different people with a whole bunch of different ideas and so it gave me a chance to look at their different ideas and see, you know, where they're coming from and they could see where I was coming from and together as a group we came up with ... exact things that could be done to fix the problems in the city. (Food and Ag Interview KI 12)

Within the process itself ... people were given opportunities to disagree in a very respectful way ... it didn't even ever feel like compromise. ... Like on that final day, everybody was happy. ... So I think the way it was structured, then the facilitation and all worked. (Food and Ag Interview KI 4)

The citizen surveys indicated an increase in the panel participants' interest in and knowledge of deliberation and local food issues over the six sessions, and in comparison to the control group. For example, the number of citizen panelists who strongly agreed or agreed on understanding the goals of the Citizen Panel rose from 79.6 per cent to 93.6 per cent over the six deliberation sessions, and the number of citizens who strongly agreed or agreed that they understood why the City of Edmonton was undertaking the citizen deliberation increased from 77.5 per cent to 97.5 per cent (Food and Ag Panel Citizen Surveys 1 and 5). With respect to cognition and behaviour change, there was some difference across time for the panelists, but it was not unidirectional or typically matched by the control group. Panelists who strongly agreed or agreed to have the city take action on reducing greenhouse gases increased from 75.5 per cent to 86.8 per cent, whereas it remained constant for the control group (Food and Ag Panel Citizen Surveys 1 and 5; Food and Ag Panel Control Group Surveys 1 and 5). The number of panelists who strongly agreed and agreed on having an interest in where the food they purchase is grown decreased from 80 per cent to 72.3 per cent, and also decreased for the control group from 71.9 per cent to 61.7 per cent (Food and Ag Panel Citizen Surveys 1 and 5; Food and Ag Panel Control Group Surveys 1 and 5). Those who felt it was very important or important that there was land for agriculture within city limits remained fairly constant over the six weeks for the citizen panelists but decreased for the control group from 72 per cent to 60.8 per cent. On the other hand, citizen panelists who indicated frequently purchasing food labelled organic increased from 22 per cent to 32.6 per cent, whereas this remained steady for the control group (Food and Ag Panel Citizen Surveys 1 and 5; Food and Ag Panel Control Group Surveys 1 and 5). The survey results are not statistically significant and therefore cannot be generalized to apply to the entire citizen panel, but they offer some indication of what I and the other researchers heard casually from the participants about the Food and Ag panel being educational and having some influence on their perceptions and behaviours.

Citizens' Panel on Edmonton's Energy and Climate Challenges

The Deliberation Design and Unique Features

The Citizens' Panel on Edmonton's Energy and Climate Challenges (Edmonton Panel) was principally a collaboration between the City of Edmonton's Office of the Environment and ABCD: fifteen members of ABCD, five members of the Office of the Environment, and two members of CPI worked together to plan and deliver the deliberation and undertake associated research, and were assisted each deliberation day by small group facilitators, note takers, and assistants (a total of twenty-three, each deliberation session) (see illustration on pp. 44–45). A highly technical Energy Transition Discussion Paper (Pembina Institute and HB Lanarc Consultants 2012), commissioned by the Office of the Environment, served as the foundational document for framing this citizen deliberation by outlining three energy scenarios: current development; reduced energy and carbon; and low energy and carbon. Citizens were directed to provide "their advice and feedback about the discussion paper recommendations: their acceptability, how far and how fast to implement them," and identify "areas of common ground and divergence" (City of Edmonton 2015).

The core planning team attempted to create a statistically representative citizen panel (MacKinnon, Dale, and Schrader 2014). The desire was to mirror the broader Edmonton population with respect to both demographic variables and attitudes on climate change, including climate skepticism or disbelief that it was human-caused, and including some people whose family incomes depended directly on the energy industry, and at least one participant from each of Edmonton's twelve municipal electoral districts (see chapter 4). While representational diversity was not fully realized, the fifty-six panelists that attended the deliberation were not the "usual suspects" who often participate in city public engagements; they reflected a wide diversity of values and perspectives. To complement this diversity of views, panelists were purposely exposed to information in a wide variety of formats over the six deliberation days (see chapter 7).

As well, a wide range of research was undertaken with groups involved. In-depth pre- and post-deliberation semi-structured interviews were conducted with the core planning team. Research to capture panelists' views and experience included pre- and post-deliberation citizen surveys, citizen journalling, and observational analysis. Post-deliberation surveys and two focus groups were conducted with the volunteer small group facilitators and note takers. As well, notes were taken during the debriefing sessions involving the core planning team and volunteer small group facilitators and note takers. Eight panelists volunteered, with the assistance of core team members from ABCD and CPI, to write the *Citizens' Panel on Edmonton's Energy and Climate Challenges Report* (CPEECC 2013), and drafts and the final version were vetted with all panelists. The Edmonton Panel defined four collective values—sustainability, equity, quality of life, and balancing individual freedom and the public good—as common ground that informed their specific recommendations; they urged City "Council and Administration to keep these [values] at the core of decision-making on energy transition issues" (CPEECC 2013, 13). The panelists recommended that the city "take the measures needed to become a low carbon city by 2050" and implement the goals and associated actions outlined in the Energy Transition Discussion Paper (CPEECC 2013, 5).

The citizens' panel report was completed January 2013 and submitted to the Office of the Environment. In April 2013, six panelists, including one person who self-declared doubt in the existence of climate change, presented their recommendations to the Executive Committee of City Council. The Office of the Environment and supportive members of Edmonton's City Council warmly received the citizen panelists' presentations, acknowledging the extended commitment shown by the citizens and how representative the Edmonton Panel was of the city as a whole. Council directed the Office of the Environment to bring back an Energy Transition Strategy based on the feedback they received. The implementation strategy came before Council in March 2015, and six members of the Edmonton Citizens' Panel attended. During the proceedings several councillors and the Office of the Environment once again drew attention to the presence of the panelists, emphasizing the representative composition of the Edmonton Panel as a whole and its resounding support for the Energy Transition Plan proposed by the Office of the Environment. The City of Edmonton's Energy Transition Strategy was approved by Council in May 2015.

CITIZENS' PANEL on EDMONTON'S ENERGY and CLIMATE CHALLENGES

1 GOAL

PROVIDE RECOMMENDATIONS TO MUNICIPAL GOVERNMENT ON HOW TO MAKE EDMONTON A MORE **ENERGY & CLIMATE RESILIENT CITY**

STRATEGIC PLAN → BUILT ON **ENVIRONMENTAL STRATEGIC PLAN** and EXPERT REPORT ON ENERGY TRANSITION

CHOOSE BETWEEN OPTIONS →
"BUSINESS AS USUAL"
"REDUCED CARBON"
"LOW CARBON"
(WITH ASSOCIATED CITY ACTIONS)

2 PLANNING with CITY of EDMONTON

2010-2012

OFFICE of ENVIRONMENT, with advice from KEY COUNCILLOR

• CENTRE for PUBLIC INVOLVEMENT (CPI)
• ACADEMICS and DELIBERATION PROFESSIONALS from ABCD

BUILDING TRUST & WORKING THROUGH MISALIGNMENTS

CO-FUNDED by CITY and ABCD

3 RECRUITMENT

DEMOGRAPHICALLY REPRESENTATIVE **SAMPLE**

INCLUDED CLIMATE SKEPTICS & ENERGY WORKERS

USED POLLING FIRM, ADMINISTERED SHORT **SURVEY**

4 DESIGN

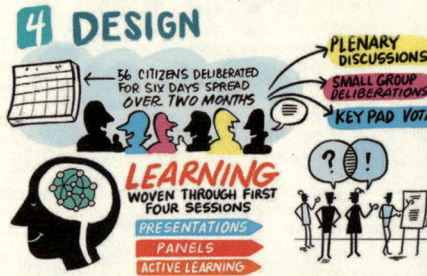

56 CITIZENS DELIBERATED FOR SIX DAYS SPREAD OVER TWO MONTHS

PLENARY DISCUSSIONS
SMALL GROUP DELIBERATIONS
KEY PAD VOTING

LEARNING WOVEN THROUGH FIRST FOUR SESSIONS
PRESENTATIONS
PANELS
ACTIVE LEARNING

5 THE SIX SESSIONS

A

MEET, LEARN and DISCUSS VALUES

B

LEARN ABOUT **CLIMATE**, DELIBERATE and **VOTE** on **GUIDING VALUES**

C

STAKEHOLDER PRESENTATIONS

DISCUSSION OF GOALS and POSSIBLE CITY ACTIONS

D

DELIBERATE ON GOALS AND ACTIONS

FIND COMMON GROUND
FIND AREAS OF DISAGREEMENT

E

DEVELOP PRELIMINARY RECOMMENDATIONS, VOTE & DEVELOP KEY MESSAGES

F

DEVELOP, VALIDATE and **VOTE** on RECOMMENDATIONS

6 OUTCOMES

- 92% SUPPORT FOR EDMONTON BECOMING LOW CARBON CITY BY 2050
- Values and Principles to GUIDE CITY WORK
- CAUTIONS, CAVEATS on PARTICULAR LINES of CITY ACTION

7 AUTHORING FINAL REPORT

SUPPORT FROM ABCD and CPI

6 PARTICIPANTS CHOSEN by PEERS TO WRITE FINAL REPORT

Validated

Validated with all PARTICIPANTS

REPORT SUBMITTED to THE OFFICE of the ENVIRONMENT

8 PRESENTATION to EXECUTIVE COMMITTEE of CITY COUNCIL
2013

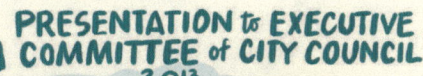

PRESENTATION

VIBRANT... CHALLENGING DISCUSSION

- 6 PANELISTS SPOKE

OFFICE OF ENVIRONMENT DIRECTED TO DEVELOP LEGISLATION

9 DEVELOPMENT OF ENERGY TRANSITION STRATEGY

ENERGY TRANSITION STRATEGY

- DONE BY CITY OF EDMONTON MANAGERS
- PANELISTS CONSULTED LIGHTLY, KEPT INFORMED

10 CITY COUNCIL PASSES ENERGY TRANSITION STRATEGY 2015

- DELIBERATES in SESSIONS ON TWO SEPARATE DAYS
- SEVERAL REFERENCES to CITIZENS' PANEL
- UNANIMOUS SUPPORT
- AND STRATEGY GETS ITS FIRST TWO YEARS OF FUNDING AT BUDGET TIME

Social Learning Outcomes

Of the four deliberations in which ABCD members participated, the Edmonton Panel represented the greatest investment with respect to both financial and time commitments. It took a year and half of negotiation and planning between ABCD and the Office of the Environment before delivering the citizen deliberation (with CPI being involved for about ten months of this), and this partnership was maintained more than two years after in order to present the citizens' panel report to Edmonton City Council and support the adoption of the Energy Transition Strategy. This sustained commitment both resulted from and deepened the relationships and feelings of trust between ABCD and the Office of the Environment (see chapter 6).

The panelists were also asked to make substantial commitments. As the Office of Environment project manager explained, we asked the citizens to do "a deep dive to understand the trade-offs associated with . . . the issue . . . . What you are doing when you are bringing people together to talk about a tough issue is that you are talking about change, and that change has a range of implications and a range of trade-offs. And so that is what we were able to do in this exercise" (Andrais 2015). As one volunteer facilitator explained, "the deliberations were so well-organized . . . they brought together people from a wide range of demographics to engage extremely complex questions thoughtfully and with mutual respect" (Edmonton Panel Small Group Facilitator and Note Taker Survey). The volume and complexity of the energy and climate challenges, and deliberation material, the technical nature of the Energy Transition Discussion Paper, and the complex policy framework translated into considerable information for many panelists to process. The lead facilitators were not trying to make the citizens into technical or policy experts but still had to work at building "citizen capacity to deal with complex issues with confidence" (KI 7-3). Most of the panelists rose to the occasion: the vast majority of panelists "came well prepared . . . motivated and ready to share ideas" (Edmonton Panel Small Group Facilitator and Note Taker Survey). As a note taker declared: "I marvelled at the commitment the panelists made and their eagerness to participate" (Edmonton Panel Small Group Facilitator and Note Taker Survey).

In addition to panelists developing increased capacity to participate, there were many indicators of instances where considerable learning was needed. One Office of the Environment staff member was frustrated when a panelist in week six of the deliberation still thought that hydro power was a source of Edmonton's electricity, when clear information had been provided to the contrary (KI 2-1).

Some panelists were able to participate more fully by the end of the six week period, learning that was observed by one facilitator:

> In the beginning there was one lady who . . . said, "You know I read all that stuff last night, three times, and I have no idea what it said." And I said, "Remember, they said that you have the ability and you have the right to just say that. And just be there." . . . I think about the fifth session I saw her and said, "Well, how are you doing?" "Good," she said. She was writing notes up for the mayor. (Edmonton Panel Small Group Facilitator and Note Taker Focus Group 2)

Likewise, in the final panel survey, when asked about what they hoped to get out of this deliberation and whether it succeeded or failed in this respect, seven panelists (n=42) talked about how much they learned. In the words of one panelist, "I learned more than I ever thought I would in six weeks and concentrated a lot more than I thought I would be able" (Edmonton Post-Deliberation Citizen Survey).

A key part of the complexity of addressing a super wicked issue like climate change in a public deliberation, especially in an energy-focused economy like Alberta's (see chapter 3), is ensuring that all views are allowed articulation, that the overall discussion is informed by science, and that collective agreement is still reached. The core team worked to ensure that there was a diversity of views in the room, and consequently some panelists strongly disagreed with one another about the energy and climate challenges facing Edmonton, yet the panelists still had to find a way to work together (see chapter 7). As one panelist explained, "Although there was [sic] disagreements the panel came to consensus on most issues and we all got to effectively participate" (Edmonton Panel Citizen Post-Deliberation Survey). This view was corroborated by several small group facilitators, as this quote demonstrates:

> One of the things I'm getting out of deliberative democracy is that people are allowed to have very opposing beliefs, and their idea doesn't have to change, they just refine it . . . . .the first day people were aggressively disagreeing with each other, but [by] the end of deliberations they were having conversations and would develop an argument. (Edmonton Panel Small Group Facilitator and Note Taker Focus Group 1)

After the sixth session, there remained a few panelists who denied the existence of climate change caused by humans, but there were indications of cognitive shifts in the group and the building of trusting relationships. While one panelist indicated in the last survey: "Now I definitely know that such [energy and climate change] issues exist and are [a] cause of a drastic change" (Edmonton Panel Post-deliberation Citizen Survey), such extreme shifts in perception about the issues were not the norm. But as this volunteer facilitator noted, there were nonetheless significant adjustments of views on the issues among the citizens:

> When you talked about the exact same thing, the exact same topic but you took a different approach to it, you used a different perspective and talked about the same thing, suddenly you had all these people who were skeptics for CC but were gung ho, on-board 100% for [putting] solar panels on every house in Edmonton because we will be more energy resilient. (Edmonton Panel Small Group Facilitator and Note Taker Focus Group 2)

As one volunteer facilitator noted, after the second week "participants [were] speaking up more comfortably but also more supportive of each other even when they disagree" (Edmonton Panel Small Group Facilitator, Note Taker and Core Team Debrief 2). It wasn't just respect that developed, but a common goal: "Most surprising was seeing that people can agree on action so closely yet have almost/seemingly opposite beliefs" (Edmonton Panel Small Group Facilitator and Note Taker Survey). From week one to week six there was a change of energy in the room as panelists became more familiar and comfortable with the process and each other, as noted by this volunteer facilitator: "There was a sense of community. People knew things about each other" (Edmonton Panel Small Group Facilitator, Note Taker and Core Team Debrief 3).

Energy Efficiency Choices Deliberation

The Deliberation Design and Unique Features

In ABCD's third year a call was sent to members of ABCD to propose deliberation projects, and Jesse Row was one of the people who responded. Row is the executive director of the stakeholder network the Alberta Energy Efficiency Alliance (AEEA) and a representative from the Pembina Institute (a Canadian environmental group). With ABCD support he hired an ABCD deliberation practitioner, Susanna Haas Lyons, to assist him in the design and delivery of an online deliberation. The purpose of the deliberation "was to engage with a representative group of Albertans ... on what government should be doing with respect to energy efficiency ... how they should fund energy efficiency programs and whether they should regulate energy efficiency standards. And to use ... what citizens think on these questions ... in the Alberta Energy Efficiency Alliance's engagement with government on the topic" (Row 2015; see illustration on pp. 50–51). At the time the deliberation was convened, the Alberta government was considering which energy efficiency programs it would support, and AEEA wanted to take advantage of this "policy window." By sharing citizens' perspectives gathered during the deliberation, the AEEA was hoping to influence government decisions about energy efficiency funding and regulations (Haas Lyons 2015).

The deliberation involved a mix of participants from across Alberta in discussions about incentives and regulations related to provincial energy efficiency programs. A professional polling firm was hired to recruit 400 Albertan participants randomly selected according to gender, age, income, education, and geography. Due to emailing difficulties and attrition, in the end only 164 citizens participated in the deliberation (see chapter 4). While diverse in some respects, the group did not mirror Alberta on several socio-demographic variables such as university education (higher than the Alberta population as a whole), and representation from those under fifty years of age (much lower than the Alberta population [Row 2014]). However, the online deliberation provided a geographically distributed discussion in which panelists entered the forum online or by telephone. Six two-hour events were held in November 2013, each with different panelists. The sessions included an orienting presentation on the topic of energy efficiency and three rounds of small group discussion in which panelists were tasked with discussing pre-established questions. The online breakout sessions were supported by volunteer small group facilitators and note takers (see chapter 7).

A participant guide developed by Row was sent to panelists in advance of the deliberation and used throughout the dialogue to help steer the discussions. The guide defined energy efficiency and outlined its economic and environmental importance, and explained that in 2010 the Alberta government allocated $30 million for advancing industrial energy efficiency but that this represents one of the lowest commitments to energy efficiency in Canada and the United States (AEEA 2013). Panelists were asked to discuss what they considered acceptable funding sources for energy efficiency programs and incentives (general revenues, GHG payments from industry, utility bills, or a new tax) and what conditions were necessary for them to support the government adopting new energy efficiency regulations. According to the lead facilitator, "it was a very instrumentally framed dialogue and not intended to explore the complexities of climate change or the complexities of our own roles or any of those things. It was more about giving advice to the government about what they can do to ensure energy use is more efficient" (Haas Lyons 2015).

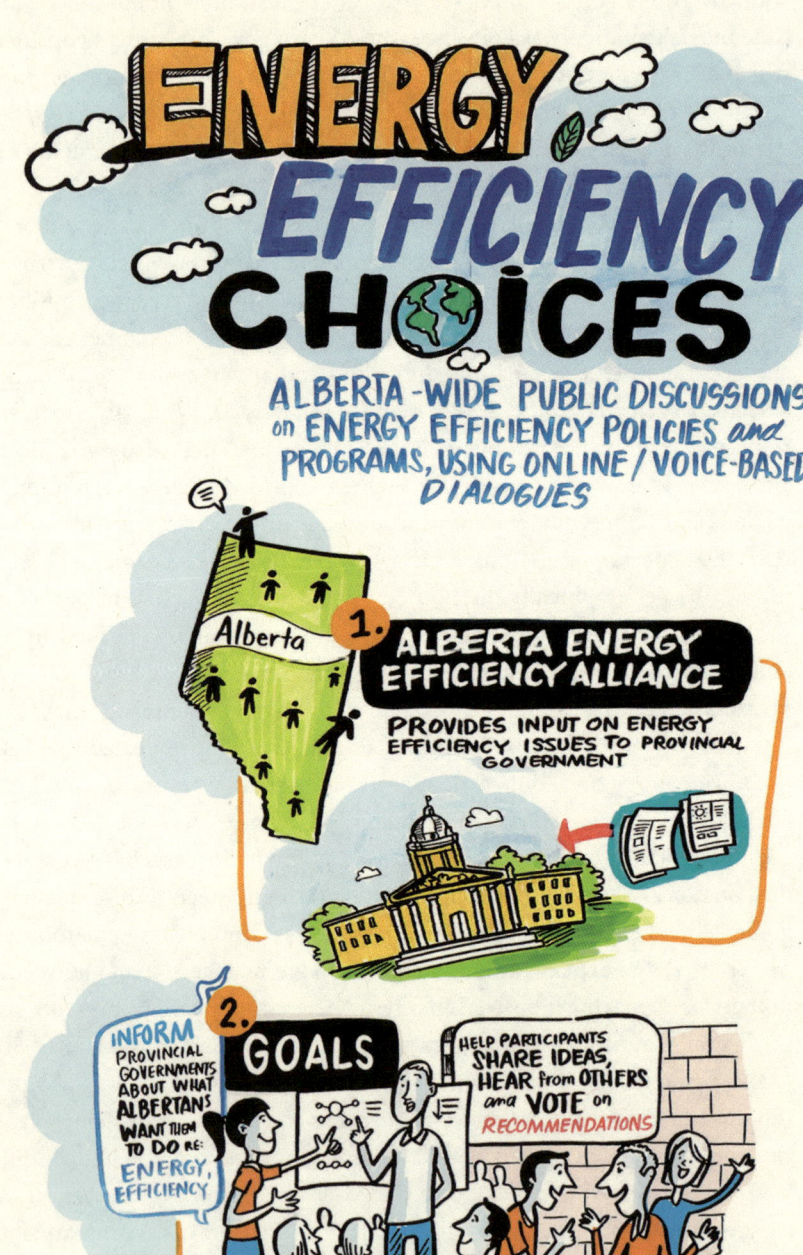

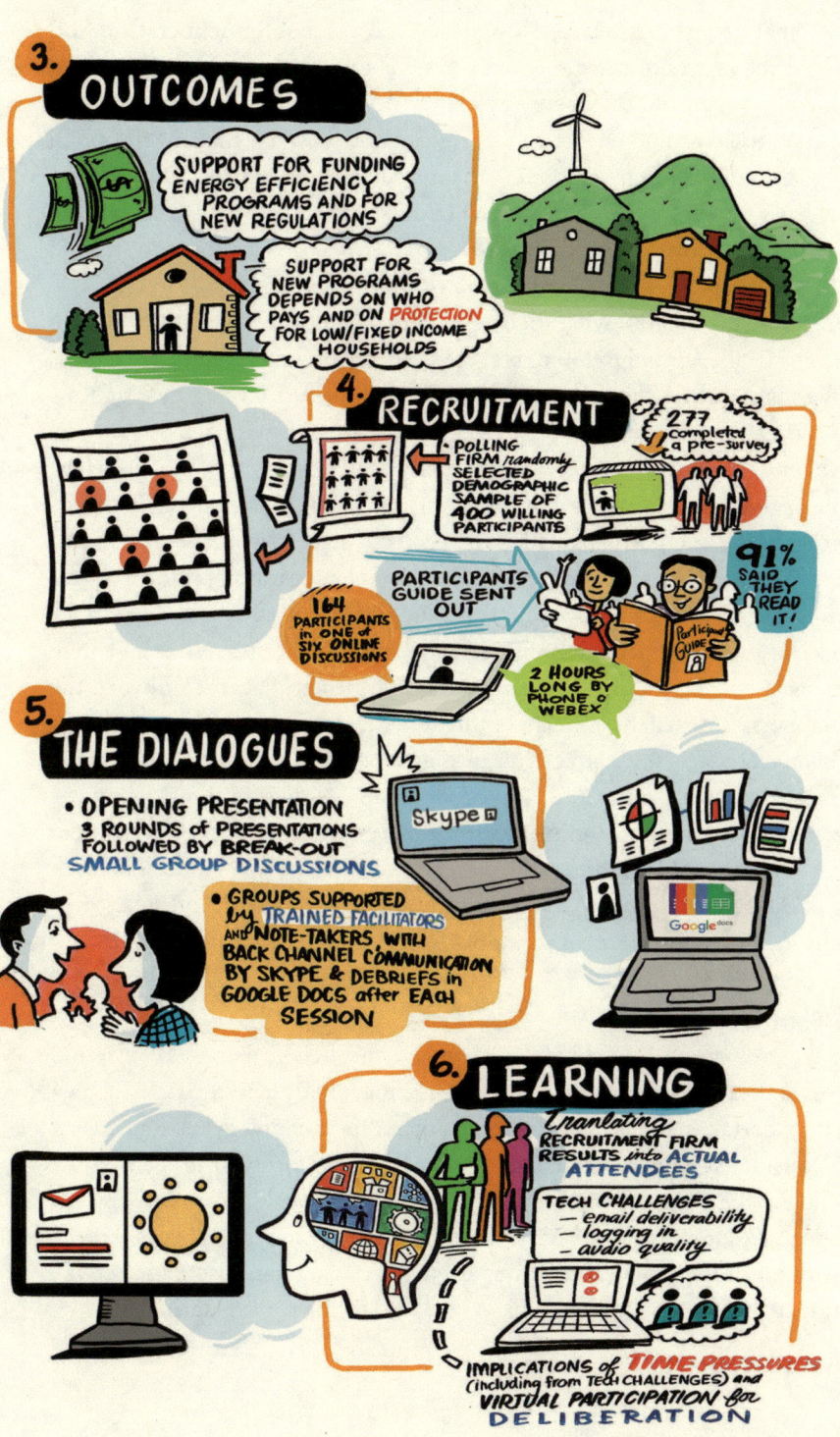

A final report written by Row described the public deliberation and detailed the results of the discussions and his follow-up survey. While ABCD financially supported the energy efficiency deliberation with an expectation that the project would enable robust research, it proved a challenge to fulfill this expectation. ABCD research was limited to short panelist surveys and interviews with Row and the lead facilitator before and after the deliberation. According to Row (2015), "the research was one of the challenges . . . faced . . . as the researchers that I had contacted didn't immediately, or weren't able to immediately identify what their research focus might be for this type of a project." In the early developmental stages of the Energy Efficiency Choices deliberation one of ABCD researchers expressed interest in interviewing Row and the lead facilitator, as well as select participants, and explained that to be useful and pertinent her interview questions would emerge from the development of the deliberation, and focus on matters such as the degree to which the organizers and participants felt the project's objectives were realized. The other ABCD researcher had developed her complete research protocol (questionnaire and email to participants) and submitted this all to Row in advance of his organizing the deliberation. Hence it wasn't that ABCD's researchers were unclear about the focus of their research but rather, that, according to the researchers, Row failed to facilitate this research because of time constraints and concerns that the publication of the ABCD research would undermine AEEA's objectives. In the end, Row conducted his own participant survey and made use of these findings in informal discussions with Government of Alberta staff. Although the influence of the citizen deliberation was likely minimal, the Government of Alberta announced new energy efficiency programs in spring 2014 (Haas Lyons 2014).

Social Learning Outcomes

The technology both enabled a distributed discussion across Alberta and impeded the full participation of panelists. In the initial planning stages, choosing to undertake a public deliberation using online technology restricted what was considered an acceptable length for the discussions and influenced the framing of the discussions: "the framing of the conversation was fairly specific because we didn't have a lot of time for meandering conversation because of the constraints of the time" (Haas Lyons 2015). The technology also impacted the quality of the panelists' discussions. As the lead facilitator explained:

The technology had an enormous impact on the deliberation . . . some persons were even unable to attend, even though they had wanted to. They'd get online, they had to figure out how it worked, even though we tried to set the bar as low as possible for technology it actually was fairly complicated for some people . . . so that impacted the diversity of the participants. The technology impacted the conversation . . . the audio quality in the small group discussion was often compromised by one individual having noise in the background and so facilitators of the small group toyed with various approaches such as having everybody muted. And when everyone was muted that took a while for their audio to kick back in . . . but then . . . they would already be talking when their audio kicked in and so people would have to stop and ask about what it was they said in the first part of their sentence. And so there was sort of an awkwardness . . . . (Haas Lyons 2015).

Many of the panelists and volunteer facilitators perceived the impacts of the technology on the discussion as negative, but certainly not all. One panelist indicated, "I think the technology somewhat got in the way of having a discussion" (Energy Efficiency Choices Post-Deliberation Citizen Survey), and a volunteer facilitator characterized the technological problems as "numerous and distracting" (Energy Efficiency Choices Small Group Facilitator and Note Taker Survey). On the other hand, another volunteer facilitator felt "the technology issues were a challenge throughout, but not enough to spoil the experience" (Energy Efficiency Choices Small Group Facilitator and Note Taker Survey). Each session, the convener and facilitator "lost anywhere from 10 to 25 minutes getting people going" (Haas Lyons 2014), and, in each of the six sessions, a few panelists had trouble connecting to a breakout group. In spite of the technological difficulties, 85 per cent of the panelists who completed the post-deliberation survey "expressed a desire to participate in this kind of thing again" (Haas Lyons 2015).

The panelists' written comments indicated mixed views regarding the educational effectiveness of the deliberation as a whole. With respect to knowledge gained, responses ranged from "[it allowed me to] learn something new" (Energy Efficiency Choices Post-Deliberation Citizen Survey) to it "didn't work for me because there wasn't enough information or purpose put forward before the actual event" (Energy Efficiency Choices Post-Deliberation Citizen Survey). Similarly, the small group facilitators and note takers, and panelist survey data, reveal that there was variable understanding of government policy processes and energy efficiency. As one panelist explained:

There were varying levels of adeptness which took up a fair bit of the time. There were also many times that we needed information that we didn't have to move our discussions forward, for example, who decides building codes and how? Industry government? Local, provincial, federal, a combination? Are they reviewed on a regular basis? etc. (Energy Efficiency Choices Post-Deliberation Citizen Survey)

The exchange of ideas and views and discussion of trade-offs that are often typical in deliberations, and that can contribute to a change of cognition and attitudes, were also limited. The breakout sessions allowed enough time for citizens to voice their views but "people did not interact with one another's opinions" (Haas Lyons 2015) because the small group discussions were so short and structured. While one panelist wrote that "it was a great way to challenge my own thoughts on energy issues" (Energy Efficiency Choices Post-Deliberation Citizen Survey), many others spoke about the lack of time to fully discuss the questions and related issues, as seen in this panelist survey response:

> Generally each individual got to make a statement and then time ran out. I had hoped to see a discussion on energy efficiency, who should be making the choices and what the consequences of those choices would be. The structure prevented any such discussion. (Energy Efficiency Choices Post-Deliberation Citizen Survey)

Perhaps most concerning was that several panelists felt manipulated by the very structured format and others spoke about not trusting that the government would take their recommendations seriously, or not feeling safe enough to fully express their views (Energy Efficiency Choices Post-Deliberation Citizen Survey). Nevertheless, approximately 80 per cent of the participants were interested in participating in another deliberation. (Boulianne and Hellstrom 2014)

Water in a Changing Climate

The Deliberation Design and Unique Features

ABCD's call for proposals in its third year also produced a deliberative partnership with Gwendolyn Blue, an academic from the University of Calgary who, like Row, had been a member of ABCD since its formation. Blue collaborated with Shannon Frank, the executive director of the Oldman Watershed Council (OWC), a stakeholder organization made up of representatives from government, industry, business, environmental groups, and ordinary citizens who provide recommendations to the Government of Alberta on how to protect and enhance the Oldman River watershed (OWC 2015). Frank remarked "that there wasn't enough community engagement around issues of climate change," and so Blue proposed to her that they collaborate on a deliberation (Blue 2015). Using funds from ABCD, Blue hired Jacquie Dale, an ABCD deliberation practitioner, to assist in designing and delivering the deliberation, as well as an outside consultant to act as the project manager (see illustration on pp. 56–57).

The one-day "Water in a Changing Climate" deliberation was held in Lethbridge, a city of just over 100,000 residents in southern Alberta. The purpose of the deliberation was to have an informed dialogue about the watershed and its future; identify common ground and public values that resonate in terms of climate change and water; and identify key topics warranting more community involvement and policy development for consideration by the OWC (Blue 2014b). Because this deliberation occurred after Edmonton's Citizens' Panel on Energy and Climate Challenges, Blue designed the deliberation as a comparative case study to engage with communities outside of Alberta's metropolitan centres, to focus on a different aspect of climate change than was the case in previous deliberations (water instead of energy), and to "see what [could] be accomplished in a day" (Blue 2015; see chapter 5).

Panelists were selected through an application process to represent diversity related to gender, age, occupation, location of residence, and views on climate change. Thirty-three participants were chosen, all of whom lived in the watershed; this included a slight over-representation of women and rural residents relative to the region's population, and also included three individuals from First Nations, making Water in a Changing Climate the only deliberation where ABCD integrated Indigenous participation into the design of the deliberation (Blue 2014a). Prior to the deliberation, all participants were sent a 26-page participant handbook. The handbook, which about half the panelists read, provided background information comparing deliberation and debate, and discussing values, climate change, and the relationship between climate change and water (Blue 2014a).

Given that there was no pressing policy decision to respond to, a flexible structure was used for the deliberation that allowed panelists to identify their own issues for discussion (Blue 2014a). The lead facilitator was assisted by five local volunteer facilitators and note takers who led and recorded the proceedings of the day's small group discussions (Blue 2014a). The morning deliberation

WATER in a Changing CLIMATE

DESIRE to TRY OUT —
→ SHORTER → 1 day PANEL to MAKE ACCESSIBLE
→ ALTERNATE FRAME - outside AB's 2 metro areas

1. OLDMAN WATERSHED COUNCIL
- MANDATE to BRING TOGETHER MULTI-STAKEHOLDERS (including citizens) TO MANAGE WATERSHED

WATER IN A 'Changing Climate'
IN TOUCH WITH OWC'ED but NO REAL DIRECTION re: OUTCOMES / INTENT

2. RECRUITMENT
- LARGELY FROM OWC LIST

DEMOGRAPHIC and SOME ATTITUDINAL DIVERSITY

REPRESENTATION from URBAN, RURAL and FIRST NATIONS COMMUNITIES

3. DISCUSSIONS
TO SHAPE PROCESS & CITIZEN HANDBOOK–

CITIZEN HANDBOOK

4. RECRUIT & TRAIN FACILITATORS
- 2 HOUR TRAINING
- OWC COMMUNITY

5. DAY

ALBERTA

Lethbridge

1. WELCOME + CONTEXT SETTING
2. PRESENTATION by DAVE SAUCHYN
3. FRAME their ISSUES & CONCERNS
4. SORTED into KEY ISSUES
5. SMALL GROUP DELIBERATION on ISSUES FRAMED by PARTICIPANTS
   - LANDUSE PRESSURES
   - EXTREME WEATHER
   - GOVERNANCE
   - SOCIAL JUSTICE/RESPONSIBILITY
   - ENVIRONMENTAL & HUMAN HEALTH
6. OWC COUNCIL & CHARGE GROUPS
7. REPORT BACK INCLUDED ACTION ITEMS
8. EVALUATION & DEBRIEF with FACILITATORS/NOTETAKERS

* filming throughout + interviews later

6. REPORT

LEAD ACADEMIC WROTE THE REPORT

- INTERVIEW of LEAD ORGANIZERS
- KEY 'Ah-hah!' FOR PARTICIPANTS = *DELIBERATION*

Ah-hah!

I GET IT!

7. OUTCOMES

- OWC SAW PANEL AS A PILOT IN OPEN-MINDED AND NEUTRAL PUBLIC ENGAGEMENT
- LEARNING ABOUT CITIZEN DELIBERATION WITH CLIMATE CHANGE – IMPORTANCE OF FRAMING, FACILITATOR TRAINING AND CAPACITY BUILDING
- MINIMAL INFLUENCE ON DECISION MAKING

began with an expert presentation on the "predicted impacts of climate change on regional water supplies" (Blue 2014a), followed by pre-assigned small group discussions focused on mapping concerns and values. The values identified included healthy environment, education, public safety, stewardship, and collective responsibility, and these values were combined with identified concerns and observations to form five themes: land use pressures, environment and public health, extreme weather events, governance, and social justice and responsibility (Blue 2014a). The afternoon activities began with a presentation by Shannon Frank about the OWC, the state of the Oldman watershed, and the kind of advice the OWC hoped to receive from the citizens' panel. Panelists self-selected themes of most interest and worked with volunteer facilitators to develop advice for the OWC on moving forward on each theme. While the morning session went smoothly, in the afternoon some of the small table facilitators "lost control of the group . . . because they didn't understand what they were being asked to do" (Blue 2015). Nonetheless, the panelists identified two key priorities: the "importance of education, information and communication," and the "significance and challenge of fostering collective responsibility for environmental protection" (Blue 2014a).

As was the case with AEEA's energy efficiency deliberation, there was some miscommunication and misalignment of expectations between Blue and ABCD's leadership and research team. In spite of there being no clear policy moment to which to tie the deliberation, ABCD leadership expected that there would be some uptake of the panelists' recommendations by the OWC, and that comparative research would be undertaken to align with the research conducted on the other three deliberations. Research activities included pre- and post-deliberation surveys of panelists, a post-deliberation small group facilitator and note takers survey, and post-deliberation semi-structured in-depth interviews with the core planning team (Blue, Dale, and Frank).

Social Learning Outcomes

A final report, *Water in a Changing Climate: Summary and Synthesis* (2014a), was written by Blue and sent to the OWC for review and feedback. The report was reviewed by Frank, but because the OWC "haven't really gotten to this issue yet about water quantity and climate change" (Frank 2015) they did not do anything with the results. However, Frank found the process educational and felt it "opened [the OWC's] eyes to a different way of approaching community discussions and not just having the old, kind of, town hall public meeting"

(Frank 2015). Likewise, 76 per cent of panelists indicated they were satisfied or somewhat satisfied with the facilitation and organization of the sessions, 82 per cent agreed or strongly agreed that the deliberation enhanced their understanding of climate change, and 79 per cent agreed or strongly agreed that it helped clarify the relevance of water to climate change (WCC Post-Deliberation Citizen Survey). When asked "what did you like best," one of the panelists wrote: "An opportunity to learn more about climate change [and] water issues from others" (WCC Post-Deliberation Citizen Survey).

While all the panelists who participated in Water in a Changing Climate were familiar with the work of the OWC to some degree, ten of the panelists felt that they were exposed to a diversity of views about climate change and water, and effective solutions to address future problems. One panelist elaborated: "I learned from peers with various backgrounds, gained new perspective, new ideas, [and] participated in the] sharing of important info" (WCC Post-Deliberation Citizen Survey). More specifically, two panelists spoke about the differences in values within the group (WCC Post-Deliberation Citizen Survey). A table host substantiated the learning that took place for some panelists, remarking: "There were a few times people said, 'oh, I did not think about that before or in that way'" (WCC Small Group Facilitator and Note Takers Survey).

While 88 per cent of the panelists indicated they were somewhat satisfied or very satisfied with the time given for discussing the issues, both Blue (2015) and Frank (2015) considered the Water in a Changing Climate deliberation to be only a good first step in engaging the public on climate change. Frank talked about the deliberation as a way to gauge where people were on the issue (2015) and Blue considered it a pilot project exploring what can be realistically achieved in one day (2015). Both Frank and Blue also said Water in a Changing Climate revealed to them the need for greater capacity than they had in terms of facilitation and organizational experience, and financial resources to carry a project like this through successfully. In Blue's words:

> It was like we were hitting the ground running with very little training for most
> of us. And there was no chance, like there was no second day where we could
> say, 'oh my goodness that went sour. What do we do about that?' ... we didn't
> forecast that outwards so we left that last bit in the air. . . Getting the report
> done, following up on how it's going to be implemented. That we didn't put into
> our charts, and as a result it kind of fell away ... these [deliberations] take a lot
> of money, a lot of resources. And they take ... a lot of capacity. (Blue 2015)

While collective action did not arise from this deliberation, nonetheless more than any of the other ABCD deliberations, Water in a Changing Climate allowed the panelists to actively set the agenda. Participatory inclusion is key to both social learning and good deliberation, serving as a stepping stone for citizens assuming a greater role in resolving public issues. Water in a Changing Climate provided a forum for the panelists to not only articulate their concerns and values related to climate change and water but also frame the problem and solutions, as well as identify how they could use their latent expertise to effect change in their community.

Discussion and Conclusion

Each of the deliberations ABCD members participated in was an experiment in how to engage citizens using deliberative methods on wicked issues. While the four deliberations had different foci, they each included some citizen engagement on climate change and, in so doing, highlighted some of the challenges in addressing the complexity of this wicked issue. On a continuum of direct citizen engagement with climate change, Water in a Changing Climate and the Edmonton Panel were most focused on the topic, whereas the Food and Ag Panel and Energy Efficiency Choices projects included only minor mention of the issue. Nonetheless, in hindsight, we in ABCD learned from each deliberation how to better engage a diverse group of citizens in a collective discussion of what climate change meant for them and what actions could effectively address it. Reflecting on how climate change was taken up in the Water in a Changing Climate deliberation, Blue (2015) explained that:

> Probably the biggest thing that was hard was the severity of the issue and how to do that, how to accommodate that . . . in a response to someone's question, [the scientific expert] said . . . if we hit tipping points we may have 30 years of a survivable climate left. And that comment never got taken up, the severity of it. I mean again, I don't think we have to believe that that comment is 100% truth, but it's a pretty significant thing, right, that the level and the depth of the problems that we potentially could be facing, I think, fell by the wayside. And people went back to a sense of safety and security and tweaking really. And so I don't know how a deliberation can be organized to help people really comprehend the potential severity of what's happening.

With its focus on values, deliberation provides a forum for citizens to make useful contributions in public policy decision-making processes, whether or not they possess a depth of scientific or technical knowledge about a topic. A focus on values is key in having people connect public policy issues to their own lives and what matters to them (Leighninger 2012). As one of the lead facilitators from the Edmonton Panel explained, including values in public deliberation directs panelists to think about the tensions and trade-offs associated with collective problem solving:

> The role of values is of course very important in deliberation . . . what kinds of trade-offs we're making with values . . . weaving back and forth. . . okay, what do these values really mean to us and are we really mindfully bringing the value choices in? And what is getting privileged and what is getting dropped? And do we understand that enough? To understand what are collective values. . . if you think about citizenship and what you are asking citizens to do as parts of a collective entity where it's not just voting or choosing for me but what does it mean to others? How values come in tension both internally and externally. (MacKinnon 2015)

Environmental management scholarship highlights the need to embed climate change adaptation in communities through, for example, having adaptation limits conceptualized within the context of thinking about how societies are organized, the values they hold, the knowledge they construct, and the relationships that exist between individuals, institutions, and the state (Collins and Ison 2009). The focus on values within deliberation provides a suitable process for better understanding stakeholder and citizen concerns and preferences in addressing climate mitigation and adaptation, as well as identifying opportunities and constraints for action at a local level. Because social learning happens "within the act and the process of constructing an issue and seeking improvements" (Collins and Ison 2009, 366), adaptive management scholarship points to the need for social learning firmly rooted in a paradigm of knowledge exchange through the emergent co-creation of knowledge, not mere knowledge transfer (McCrum, et al. 2009). ABCD's experiences with public deliberation demonstrate that citizens are able and willing to act in the collective interest, and that when enabled and supported, they can make important contributions in identifying both the nature of the problem and possible solutions.

Social learning refers to the individual learning that is conditioned by the social environment, as well as learning by social collectives such as organizations (Pelling et al. 2008). This chapter has focused on the first of these in order to highlight views and experiences of the citizens who were central to this project and each deliberation. But in planning to address wicked issues, it is crucial that both forms of social learning are recognized and built into the process, as they are complementary. "In turn this requires new roles and practices relating to facilitation and new kinds of institution and policy" (Collins and Ison 2009, 367), which, as mentioned previously, speaks to the need for more flexible, inclusive, and iterative processes, policies, and institutional contexts. The transformations in democracy, society, and public engagement necessary to address the complexity, uncertainty, immediacy, and multiple stakeholding associated with wicked issues are formidable but not impossible. As these deliberation profiles illustrate, it is not easy to convene deliberations that fully engage citizens in decisions about complex topics like climate change adaptation and mitigation; none of the ABCD deliberations fully explored the multiple layers of social learning required to address climate change, in part because of our own shortcomings, but also because the existing regulatory, fiscal, and educative approaches are insufficiently capable of accommodating this kind of praxis. But in reflecting back on these projects, we are able to identify ways to increase social learning.

The remaining chapters of this book look at these issues in some detail, with an eye to answering the question: How can we use deliberation to better address the complexity, uncertainty, immediacy, and multiple tensions associated with wicked issues?

Note

1. The interview data quoted in this chapter comes from four sources: in-depth semi-structured interviews conducted for each of the four deliberations, mostly of core planning team members; small group facilitator and note taker surveys undertaken for each deliberation; citizen surveys, often pre- and post-deliberation, but, for the Food and Ag Panel, throughout the deliberation and matched by control group surveys; and notes from debriefing sessions for the Edmonton Panel with small group facilitators, note takers and the core planning team. Some of the individuals interviewed are identified by name in the chapter, with their informed consent. For many others, their identity remains confidential, and to anonymize them, the deliberation or policy process is identified in the text, and in an interview a number given to distinguish the key informant, and if applicable, this is followed by a number to identify whether the interview was pre-deliberation (1) or post-deliberation (2). If the comments arose out of a focus

group or from a survey, the kind of participants are identified (e.g., small group facilitators and note takers) but no numbers are assigned to distinguish individual focus group participants' comments.

References

AEEA (Alberta Energy Efficiency Alliance). 2013. "Energy Efficiency Choices Participant Guide: A Citizens' Discussion on Alberta's Energy Efficiency Policies and Programs." Unpublished Deliberation Participant Guide.

Andrais, Jim. 2015. Telephone Interview with Lorelei Hanson, December 9.

Bandura, Albert. 1977. *Social Learning Theory*. Englewood Cliffs, NJ: Prentice Hall.

Baron, Robert S., and Norbert L. Kerr. 2003. *Group Process, Group Decision and Group Action*. Milton Keynes: Open University Press.

Beckie, Mary A., Lorelei L. Hanson, and Deborah Schrader. 2013. "Farms or Freeways? Citizen Engagement and Municipal Governance in Edmonton's Food and Agriculture Strategy Development." *Journal of Agriculture, Food Systems, and Community Development* 4(1): 15–31. doi:10.5304/jafscd.2013.041.004.

Blue, Gwendolyn. 2014a. *Water in a Changing Climate: Summary and Synthesis*. Unpublished Report.

———. 2014b. "Water in a Changing Climate Citizen Deliberation Debrief Meeting Background Notes." Unpublished Internal ABCD document. March 12.

———. 2015. Telephone Interview with Lorelei Hanson, June 8.

Boulianne, Shelley, and Mikael Hellstrom. 2014. "Citizen Perceptions of the Efficacy of Deliberative Exercises." Unpublished ABCD research brief.

Cavanagh, Fiona. 2015. Telephone interview with Kristjana Loptson, June 5.

Centre for Public Involvement. 2012a. *City-Wide Food and Urban Agriculture Strategy Report on Citizen Panel Process and Recommendations*. July 26. http://centreforpublicinvolvement.com/work/archives/2012/11/28/the-report-on-citizen-panel-process-recommendations/.

———. 2012b. "Session Outputs and Schedule." Unpublished Internal Document. April 15.

City of Edmonton. 2006. Public Involvement Policy, no. C513. January 17. http://www.edmonton.ca/transportation/C513.pdf.

———. 2012. *fresh: Edmonton's Food and Urban Agriculture Strategy*. http://www.edmonton.ca/City_government/documents/FRESH_October_2012.pdf.

———. 2015. Citizens' Panel on Edmonton's Energy and Climate Challenges. http://www.edmonton.ca/city_government/environmental_stewardship/citizens-panel-energy-climate.aspx.

Collins, Kevin, and Ray Ison. 2009. "Jumping off Arnstein's Ladder: Social Learning as a New Policy Paradigm for Climate Change Adaptation." *Environmental Policy and Governance* 19: 358–73.

CPEECC (Citizens' Panel on Edmonton's Energy and Climate Challenges). 2013. *Citizens' Panel on Edmonton's Energy and Climate Challenges Report* https://www.edmonton.ca/city_government/documents/PDF/CitizensPanel-EnergyClimateChallenge.pdf.

Diduck, Alan, A. John Sinclair, Glen Hostetler, and Patricia Fitzpatrick. 2012. "Transformative Learning Theory, Public Involvement, and Natural Resource and Environmental Management." *Journal of Environmental Planning and Management* 55(10): 1311–30.

Frank, Shannon. 2015. Telephone Interview with Lorelei Hanson, July 19.

Haas Lyons, Susanna. 2014. "Learning from Technology's Role in Energy Efficiency Choices." Unpublished Alberta Climate Dialogue Workshop Paper.

———. 2015. Telephone Interview with Lorelei Hanson, May 14.

Hanson, Lorelei L., and Deborah Schrader. 2014. "Creating New Urban Spaces of Sustainability and Governmentality: An Assessment of the Development of a Food and Urban Agriculture Strategy for Edmonton, Canada." In *Sustainable Cities: Global Concerns/Urban Efforts*, edited by William G. Holt, 195–218. Bingley, UK: Emerald Group.

HB Lanarc Consultants. 2012. Agricultural Inventory and Assessment: City of Edmonton City Wide Food and Agriculture Strategy. Edmonton: City of Edmonton.

Johnson, Genevieve Fuji. 2009. "Deliberative Democratic Practice in Canada: An Analysis of Institutional Empowerment in Three Cases." *Canadian Journal of Political Science* 42(3): 679–703.

Kay, James J., and Eric Schneider. 1994. "Embracing Complexity: The Challenge of the Ecosystem." *Alternatives* 20(3): 32–39.

Lave, Jean, and Etienne Wenger. 1991. *Situated Learning: Legitimate Peripheral Participation*. Cambridge: Cambridge University Press.

Lazarus, Richard J. 2009. "Super Wicked Problems and Climate Change: Restraining the Present to Liberate the Future." *Environmental Law and Policy Annual Review* 40: 10749–56.

Leighninger, Matt. 2012. "The Next Form of Democracy?" Civic Engagement and Democracy Lecture, Institute for Policy and Civic Engagement, University of Illinois, Chicago, April 4. http://www.deliberative-democracy.net/index.php/resources/guides-and-reports.

Levin, Kelly, Benjamin Cashore, Steven Bernstein, and Graeme Auld. 2012. "Overcoming the Tragedy of Super Wicked Problems: Constraining Our Future Selves to Ameliorate Global Climate Change." *Policy Sciences* 45(2): 123–52.

MacKinnon, Mary Pat. 2015. Interview with Lorelei Hanson, May 7.

MacKinnon, Mary Pat, Jacquie Dale, and Deborah Schrader. 2014. "Looking Under the Hood of Citizen Engagement: The Citizens' Panel on Edmonton's Energy and

Climate Challenges, ABCD." Working Paper. http://www.albertaclimatedialogue.ca/wp-content/uploads/2016/02/ABCD_WP_PractitionersReflecitons_2014-09-02.pdf.

McCrum, Gillian, Kristy Blackstock, Keith Matthews, Mike Rivington, Dave Miller, and Kevin Buchan. 2009. "Adapting to Climate Change in Land Management: The Role of Deliberative Workshops in Enhancing Social Learning." *Environmental Policy and Governance* 19(6): 413–26.

Muro, Melanie, and Paul Jeffrey. 2008. "A Critical Review of the Theory and Application of Social Learning in Participatory Natural Resource Management Processes." *Journal of Environmental Planning and Management* 51(3): 325–44.

Nutter, Monique, Debbie Hubbard, and Butch Nutter. 2011. "Food Security in Edmonton's Municipal Development Plan." Presentation given at the Faculty of Social Work Research Day. Calgary: University of Calgary.

OWC (Oldman Watershed Council). 2015. "Vision Mission and Goals." Accessed February 2. http://oldmanwatershed.ca/visionmissiongoals/.

Pahl-Wostl, Claudia, Jan Sendzimir, Paul Jeffrey, Jeroen Aerts, Ger Berkamp, and Katharine Cross. 2007. "Managing Change toward Adaptive Water Management through Social Learning." *Ecology and Society* 12(2): art. 30. http://www.ecologyandsociety.org/vol12/iss2/art30/.

Pelling, Mark, Chris High, John Dearing, and Denis Smith. 2008. "Shadow Spaces for Social Learning: A Relational Understanding of Adaptive Capacity to Climate Change within Organizations." *Environment and Planning A* 40: 867–84. doi:10.1068/a39148.

Pembina Institute and HB Lanarc. 2012. Edmonton's Energy Transition: Discussion Paper. https://www.edmonton.ca/city_government/documents/PDF/Edmonton_Energy_Transition_Discussion_Paper.pdf.

Reed, Mark S. 2008. "Stakeholder Participation for Environmental Management: A Literature Review." *Biological Conservation* 141(10): 2417–31.

Row, Jesse. 2014. "Energy Efficiency Choices: Report on Citizens' Views." Unpublished Report.

———. 2015. Telephone Interview with Lorelei Hanson, Deborah Schrader, Mary Pat McKinnon, and David Kahane, June 7.

Tompkins, Emma L., and W. Niel Adger. 2004. "Does Adaptive Management of Natural Resources Enhance Resilience to Climate Change?" *Ecology and Society* 9(2): 10. http://www.ecologyandsociety.org/vol9/iss2/art10.

Torres Scott, Andre. 2012. "Report on the Recruiting Phase, Food and Agriculture Panel." Unpublished Internal Centre for Public Involvement Document.

Waltner-Toews, Walter, James J. Kay, Cynthia Neudoerffer, and Thomas Gitau. 2003. "Perspective Changes Everything: Managing Ecosystems from the Inside Out." *Frontiers in Ecology and the Environment* 1(1): 23–30.

Wilner, Kate Bigney, Melanie Wiber, Anthony Charles, John Kearney, Melissa Landry, Lisette Wilson, and on Behalf of the Coastal CURA Team. 2012. "Transformative Learning for Better Resource Management: The Role of Critical Reflection." *Journal of Environmental Planning and Management* 55(10): 1331–4. doi:10.1080/09640568.2011.646679.

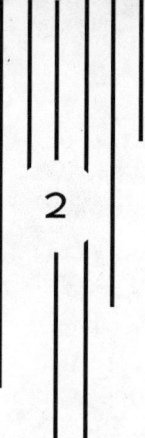

2

The Theory and Practice of Deliberative Democracy

David Kahane and Gwendolyn Blue

Mitigating and adapting to climate change are among the most pressing challenges we face. The ability of humans to respond wisely, effectively, and quickly will determine the future of our species and of the planet we share with others. This volume as a whole looks at how citizens of Alberta took part in four deliberations related to climate change, and at the difference these deliberations made in a province with an economy driven by fossil fuel extraction. Before Alberta Climate Dialogue (ABCD) came into being in 2010, the Government of Alberta had already engaged citizens in public consultations on climate change, but in ways that were cursory and had little apparent influence (Adkin et al. 2016). The introduction to this volume provided an overview of climate change and climate politics, and of the evolution of deliberative democracy as a field of theory and practice. This chapter digs more deeply into debates within deliberative democracy and how they found expression in ABCD; and it explores the challenge, promise, and potential pitfalls of bringing the tools and frameworks of deliberative democracy to debates and politics around climate change, and how these played out in ABCD.[1]

How ABCD Negotiated Debates around Deliberative Democracy

Deliberative democracy, or public deliberation, is one form of citizen participation, alongside a diversity of other approaches. Public deliberation has four characteristics that distinguish it from other discourses and practices of engagement, involvement, participation, and consultation. Firstly, an emphasis on representing the diversity of affected communities in political discussion and problem solving, as distinct from approaches that throw open the doors of a public engagement process and are satisfied with whoever shows up (on this spectrum see Lukensmeyer 2012). Secondly, an emphasis on deliberation: the view that well-designed and effectively implemented processes not only elicit perspectives from participants but give them the information and learning opportunities they need to ground their perspectives, involve them in a careful back-and-forth with those holding different perspectives, and support them in weighing complex considerations and trade-offs (Bohman 1996). Thirdly, deliberative democracy emphasizes the importance of real influence, and of collective decision making in light of this influence: participants should understand themselves as able to affect political outcomes, and should develop common recommendations and/or plan actions in this light. And fourthly, deliberation should be rooted in participants' values, should support reflection on values, and should orient participants to the possibility of articulating shared, common, or civic values as a basis for their deliberation and decision making (Lukensmeyer 2012).

These four characteristics should not be taken to suggest that deliberative democrats line up around some shared view of the field: there is much debate in both research and practice. Eight key areas of divergence and debate within deliberative democracy were very much alive within ABCD as well.

A first area of divergence within deliberative democracy theory and practice is between *claimed* democratic spaces, which build from the grassroots up, and *invited* spaces that originate from a government or other organization and reach out to engage more broadly (Gaventa 2006). Different theories and practices tend to foreground claimed or invited spaces, and to read such spaces through different accounts of social and political change. This is not a binary choice: projects often combine claimed and invited aspects, and indeed, the influence of a process can depend on both having roots in community self-organization and allies within powerful organizations or government (Gaventa and Barrett 2010). ABCD began with an emphasis on partnership with government to hold deliberative processes, but it went on to work with civil society and para-governmental organizations as conveners (see Chapter 6 for further discussion of this evolution). All of ABCD's deliberations were invited rather than claimed spaces.

A second area of debate in deliberative democracy circles is around the influence that processes need in order to be empowering, effective, and legitimate. Many public consultations run by governments and other organizations seek to elicit the views of citizens, but the voices elicited have obscure or minimal influence on decision making. Some deliberative democrats are willing to work with this sort of constrained influence (Lee 2015). Others see publicity, transparency, and strong public profile for deliberative exercises as a linchpin of influence (Fishkin 1997). And others still see formal commitment to decision-making influence as a principle-driven requirement of good deliberative processes, as well as instrumental in persuading diverse citizens to take part (Fung 2003). ABCD, from its beginnings, saw influence on governments and policy as key in the Alberta context, and oriented its efforts to developing partnerships with government decision makers who would commit to giving uptake to outcomes of citizen processes. This focus was established at a workshop of researchers and practitioners convened in 2009 to put together the application for the Canadian government grant that eventually funded the project. Provincial policy was seen as key to addressing challenges of climate mitigation, and thus as a key target for citizen recommendations; at the same time, there was some pessimism about whether the provincial government would be willing to partner or meaningfully incorporate citizen recommendations, and so municipal governments were taken to be key interlocutors. Recruiting municipal partners proved a steep hill to climb in the Alberta context, though it succeeded with the City of Edmonton (see chapters 6 and 7). Moreover, some members of ABCD were interested in other potentials of citizen deliberation: the Energy Efficiency Choices process explored how deliberation can support NGO lobbying efforts, and Water in a Changing Climate aimed to broaden the frame to include adaptation and water, and was regarded by the Oldman Watershed Council as a way to learn about public deliberation, not a direct input into decisions (see below and chapter 5).

A third area of divergence and ferment in deliberative democratic theory as well as practice involves the distinction between designing one-off deliberative forums and building deliberative systems. Much work in public consultation, engagement, involvement, and deliberation builds contexts for citizen deliberation that have a clear beginning and end in terms of organization, funding, professional support, and outcomes that feed determinately into a decision-making process. There is increasing emphasis among deliberative democrats, though, on both understanding punctual exercises against a backdrop of more complex ecosystems of deliberative and democratic settings and institutions in a community (Chilvers and Longhurst 2012; Mansbridge et al. 2012) and on building capacity for communities, civil society organizations, and governments to go beyond one-off engagement to create the cultures and institutions of a deliberative society (Nabatchi and Leighninger 2015; Mansbridge et al. 2012). All of ABCD's deliberations were one-off mini-publics, though there was a lot of discussion and effort devoted to understanding these deliberations in their richer contexts, as well as to rooting them in civil society organizations and mobilization. This was particularly true of the Citizens' Panel on Edmonton's Energy and Climate Challenges: early proposals from ABCD sought to weave together a mini-public with community-based projects involving both panelists and civil society organizations, though the eventual project did not include this civil society dimension (see chapter 6).

A fourth debate in deliberative democracy, especially in theory, concerns the forms of reason and narrative that support good deliberation. Deliberative democratic theory has some of its roots in approaches that emphasize the importance of adhering to principles of reasoned agreement in deliberations, as distinct from using evocative narratives or passionate rhetoric to sway discussions (Habermas 1993; see also the introduction and chapter 7). Critics of this emphasis on reasonable agreement point to how norms of reasonableness often are used to marginalize certain groups, including women, colonized peoples, people of colour, and others who face or have faced systematic inequality and exclusion from scientific, technical, and philosophical institutions. Norms of reasonableness are not uniform across social groups. These critics describe how legitimate deliberation and joint decision making can take place using more diverse forms of expression (Young 2005; Williams 1998). In ABCD, we sought to make a place for narratives and situated forms of understanding in our processes, but we also wrestled with the privileging of certain forms of expert reasoning and knowledge over others, particularly in ways of communicating climate science and technologies for energy transition. For example, climate scientists as well as civil servants with engineering backgrounds were prominent communicators in the learning stages of both the Oldman Watershed and Edmonton deliberations, and their perspectives were likely accorded weight by participants because of their scientific credentials, as well as assumptions carried by participants about climate change as a complex scientific issue. Attending to how certain groups and sources of expertise are treated as authoritative in speaking about climate change presented a challenge in our collective deliberations (see Blue 2015; Blue and Dale 2016, for an extended discussion).

Related to this debate, a fifth area of discussion within deliberative democratic theory and practice has to do with framing issues for deliberation. Every discussion has a frame: a set of terms, constraints, and assumptions that provide a starting point, a set of tendencies, and potentially a set of formal limitations on debate. Deliberative democrats tend to emphasize the importance of working carefully with diverse groups to define appropriate and legitimate frames for deliberation; being explicit with participants about frames, as well as what is formally on the table and what's not; and giving scope to participants to challenge frames (Kettering Foundation 2011). But there is a range of views on how these kinds of principled commitments deal with power relations in determining frames; with dominant and marginalized discourses as they shape frames and how conveners, facilitators, and participants engage with frames; and with how norms of reasonableness shape framing (Ulrich 2005). These discussions were alive in ABCD, and we repeatedly experienced how a principled commitment to sensitivity to power relations in framing came up against the reality that dominant frames for climate change and for citizen consultation repeatedly reasserted themselves, and often held sway within our work (see chapters 5 and 7). For example, scientific and technological framings of climate responses were central to the Edmonton deliberation, given the policy moment that deliberation was meant to address, the assumptions that participants brought into the room, the materials provided to participants, and the backgrounds and perspectives of resource people we brought into the room to answer questions. Inequalities of power and framing are challenging to address in practice because they reflect broad systemic tendencies, ideologies, and unquestioned assumptions; these dynamics are slow to change, and are not typically able to be resolved by finding the "right" deliberative procedure. Given this, it is important to sustain acknowledgment and reflection about the powerful role that social context plays in the design and conduct of deliberative initiatives, and to push back against understandings of these spaces as unproblematically neutral.

These power and framing inequities tie into a sixth area of debate within deliberative democracy, the relationship between activism and deliberation. Some approaches privilege public deliberation as a legitimate way of addressing public disputes: different parties bring their claims and perspectives into well-designed forums, which adjudicate between these in terms of the public good. Others acknowledge that where there are unjust forms of power and exclusion, activism can be necessary to bring issues to the deliberative table, and to ensure that they are taken up fairly (Fung 2005). And still others are unwilling to privilege deliberation over activism, and insist that both have their role within a healthy deliberative system (Cornwall and Gaventa 2001). Many members of ABCD were connected to the worlds of activism and climate activism; they wanted deliberations to connect to the energies already present in civil society around climate issues. One complicating factor was the political culture and history of Alberta, where climate change was highly politicized and publics were less mobilized on climate issues than in many other contexts (see chapter 3). The interest of members of ABCD in connecting to activist politics was also complicated by the political situations and strategic interests of partners in deliberation: the City of Edmonton and the Alberta Energy Efficiency Alliance wanted their public deliberations to be broadly seen as legitimate, and to them this entailed limiting the profile of activist languages and groups in the design, materials, and profile of each exercise.

Seventh, there are divergences of practice, and to some extent of theory, around how to understand the relationship between processes that engage citizens and processes that engage organized stakeholder groups (Kahane et al. 2013). Each of ABCD's deliberations engaged individual citizens, recruited in ways that sought to ensure participant diversity; but each of these processes sat alongside others that engaged organized stakeholders—for example, from the private sector, civil society, and governments. The City of Edmonton, for example, had involved stakeholder representatives extensively in developing *The Way We Green*, the environmental strategic plan that created the context and political moment for the Citizens' Panel on Edmonton Energy and Climate Challenges, and civil servants worked with stakeholder representatives in developing the energy pathways between which panelists chose, and in developing an Energy Transition Strategy based in part on panel recommendations. There was minimal crossover between these citizen and stakeholder processes, and opacity around how these different inputs would shape city policy development. This separation of stakeholder and citizen tracks is typical of public deliberation work, though some deliberative approaches bring organized stakeholders and individual citizens into a common deliberative space, or keep deliberations separate but explicitly cross-fertilize them (Kahane et al. 2013).

Eighth, and finally, there is a new current of debate in scholarly analyses of deliberation about the professionalization of public deliberation. Proponents of the professionalization thesis suggest that while the self-image of practitioners of democratic deliberation and other approaches to public participation tends to emphasize the diversity of methods employed, the principles behind the methods, and the progressive democratic outcomes of these processes, the field is in fact organized in ways that tend to deliver quite uniform processes that conform to the interests of the powerful, with ambiguous democratic outcomes (Lee 2015; Lee, McQuarrie, and Walker 2015). For instance, deliberative formats tend to follow quite similar formats such as the use of roundtables, initial warm-up exercises and discussions about core values and concerns, breakout sessions, return to large group with report-backs, and process summaries. This standardization is not a problem in itself and it occurs in any professional field. But insofar as claims about the sameness of deliberative methods are warranted (and it can be hard to sort out which forms of regularity or divergence are most significant in such a sprawling domain of practice), they give reason to temper claims about design innovation. ABCD's work, and internal conversations, took up this critical concern about serving dominant interests through regularized practices in some moments, and in other moments the concern was subsumed in the hard work of getting deliberations done, in the positive rhetoric of the transformative potential of deliberative processes, and in the financial imperatives that condition the work of participation professionals (Blue and Dale 2016).

These eight debates about deliberative democracy were a subtext to ABCD's work to convene public deliberations on climate change in Alberta. They emerged repeatedly in internal ABCD meetings and discussions, particularly at the three major team workshops in 2010, 2011, and 2014. And they inflected the development of particular partnerships and deliberation projects. As important, though, and as troubling at times, was how the tools (and debates) of deliberative democracy related to the distinctive challenges of climate change. In the next section, we look at how public deliberation encounters particular challenges, and holds particular potentials, when it comes to climate change.

The Challenge of Addressing Climate Change Through Public Deliberation

A recurrent challenge of developing robust climate policy is linking the scientific consensus on climate change with a concerted political effort about what to do in response. One view holds that scientific agreement should, in principle, facilitate concerted action by providing a baseline of shared information for all to follow. Most credible scientists agree that the Earth's global temperature is rising and that climates are changing as a result of collective human activity. Furthermore, it is well documented that those scientists who challenge this consensus have ties to corporate and political interests that seek to maintain the profitability of fossil fuel industries (Oreskes and Conway 2010). Yet the scientific consensus on climate change—developed and communicated by global knowledge assessment institutions such as the IPCC—has not resulted in concerted political action and concern on the level many argue is needed to prevent dangerous climate change by the end of the twenty-first century.

For some commentators, the problem lies with ignorant and easily duped publics and policy makers who lack scientific literacy and therefore need to be better informed about the issue. This position has led to efforts to communicate climate science more straightforwardly to various constituencies. These efforts have not proven effective, in large part because of unexamined and questionable assumptions about public deficits of knowledge. As Susanne Moser and Lisa Dilling argue, a more comprehensive approach to political communication is warranted: "People in a democratic society are best served by actively engaging with an issue, making their voices and values heard, and contributing to the formulation of societal responses. . . . Effective communication serves two-way engagement, which—ultimately—enables societal action" (2010, 169).

Climate change is a complex issue with many different definitions and approaches. While current scientific consensus states that humans are influencing the climate, there remains much disagreement over how to make sense of climate change and what to do in response. Moreover, climate change is not only a scientific issue but also a deeply cultural, political, and ethical one (Hulme 2009). Significant disagreements exist among researchers across the sciences, social sciences, and humanities, as well as among activists and policy makers, about how to best interpret and act on climate change. These differences provide a much richer terrain of interpretation than is typically captured by attempts to divide the world into "believers" and "deniers."

Deliberative democracy provides an attractive option for public engagement. The general agreement across varied positions is that citizens should have the right to voice their values and perspectives about climate policy and that policy makers need to defend their political decisions to those whose lives are affected by them. The deliberative turn in democratic theory highlights the significance of communication and reflection in political processes, so that democracy is not seen as simply the aggregation of preferences in order to inform decision making but is "also about processes of judgment and preference formation and transformation within informed, respectful and competent dialogue" (Dryzek 2006, 3).

Dryzek outlines why deliberative democracy can help environmental decision making in general and climate change in particular (2013, 13). First, deliberation can help integrate different perspectives. Second, the kinds of values that emerge from a deliberative setting can assist with prioritizing collective interests over material self-interest. Third, it can enable new ethical relations to emerge in that it expands the thinking of its participants to better encompass the interests of future generations, distant others, and non-human nature. Fourth, it can organize feedback on the condition of social-ecological systems into politics.

While this overview offers important links between deliberation and climate change, significant challenges arise in practical settings that complicate these aspirations. As noted, public deliberation differs from other forms of democratic engagement in that, ideally, it is inclusive of diverse communities and perspectives; in practice this is challenging. Some researchers argue that diversity, inclusion, and equity are not in fact central concerns for deliberative advocates and practitioners (Lee 2011), and we can see several reasons for this. As Leighninger (2010) describes, one explanation is that deliberative advocates tend to be mostly white and from relatively socially privileged groups, a point that was by and large true of the ABCD team. As discussed in chapter 6, ABCD struggled to break out of this circle, but with limited success. For example, repeated attempts in the first two years of the project to involve Indigenous people and communities were limited in their success. The reasons for this are not well understood. Given that climate change impacts social groups differently, questions of who has the privilege to frame debates and democratic processes are significant, and early decisions about subject matter and framing may themselves constrain which groups are interested in investing limited time and resources in deliberation projects.

As well, the tendency to focus on questions of process, design, and impact on policy can depoliticize public deliberation, including by downplaying structural hierarchies based on gender, class, and race (Lee 2011; Hendriks and Carson 2008). In convening citizen panels, a lot of ABCD's efforts around social diversity related to recruiting for demographic representativeness (see chapter 4), though inclusion and representativeness are not necessarily equivalent.

Addressing the relationship between representativeness and inclusion is challenging in practice. Analyses of group-based hierarchy, inequality, and domination are often hard-won results of collective organizing and activism; addressing these within specific deliberations requires an engagement with the political character of group memberships, and the need to attend to and perhaps amplify the voices of marginalized groups or groups disproportionately affected by an issue. While individuals from particular groups may be important in bringing in group-based perspectives, formulating, articulating, and securing uptake for group-based perspectives is an importantly collective project; it may require that the design of deliberation support solidarity, collective analysis, and collective voice by members of marginalized communities, through mechanisms like oversampling in recruitment, creating separate spaces of deliberation for particular groups, and supporting caucusing within deliberations by particular groups (Kahane 2002; Blue, Medlock, and Einseidel 2012).

Recognition of these challenges does not diminish the significance of democratic representativeness. The demographic diversity of ABCD's deliberations was important, and did shape outcomes. In the Citizens' Panel on Edmonton's Energy and Climate Challenges, for example, the class diversity of those in the room helped to bring issues of economic disadvantage and neighbourhood violence into discussion, and influenced recommendations, for example, around the need to attend to equity and public safety in supporting mixed-use, transit-oriented neighbourhoods. Yet bigger questions of structural inequality might have been addressed quite differently had we taken an alternative approach to group representation from the outset.

Another challenge is that emphasis on gearing deliberation toward policy and government decision makers meant that certain types of questions and policy responses could get framed out of discussion (see chapter 7). For example, although certain activist communities have tried to question whether existing political and economic structures (capitalism, for example) are contributing to climate change, these avenues of inquiry are not typically entertained by policy makers (and if they do entertain them, it can be politically dangerous for them to express such views publicly or professionally). During the time in which ABCD was active, groups like the Indigenous Environmental Network and scholars like Ian Angus actively raised questions about the relationship between climate change and existing economic structures. These more "radical" ideas were absent from or at the margins of educational materials and framing in ABCD's deliberations, and invited experts and civil society speakers at the citizen panels did not represent such positions. Even where issues like the unequal distribution of carbon emissions or climate impacts globally were brought into educational materials—in the Citizens' Handbook for the Edmonton Climate Panel, for example—they were not foregrounded in the carbon scenarios that framed the main choices made by panelists. Participants were free to bring alternative policy positions into deliberation, and some did (at the Citizens' Panel on Edmonton's Energy and Climate Challenges, for example). But the salience of alternative positions was limited by framing and by the choices foregrounded in the design of trajectories of conversation.

Geared toward pragmatic policy ends, formalized public deliberation can reinforce rather than question existing social structures, a point that has long been argued by critical theorists. Addressing climate change might well require fundamentally questioning existing social and political systems and examining trade-offs and opportunities presented by alternative policy proposals. Broader contexts of social power are often sidelined or naturalized in the design and conduct of deliberative exercises like those held by ABCD, and this has direct implications for how climate change is framed and on what political responses are treated as most salient.

Responding adequately to questions of social power in designing climate deliberations would, as the discussion above brings out, require a more politicized approach to group representation; more grassroots-up processes of framing; a different composition of project teams from an early stage; and strategies of influence not premised on partnership with policy makers, who often will constrain themselves (consciously as well as unconsciously) to pragmatic, mainstream solutions as opposed to the tangled and difficult work of raising critical questions about framing, influence, and power. There is a need for experimentation, innovation, and research on these alternative approaches, which can help to foster social learning on the part of policy makers, scientific experts, academics, and deliberation practitioners on their institutional and cultural biases and assumptions. Such reflexive self-questioning is challenging in practice and can encounter much resistance (Pallett and Chilvers 2013).

These challenges do not diminish the importance of public deliberation, but do signal its potential limits. Policy reform within existing social structures is important for addressing climate change, particularly as changes to existing laws, regulations, and institutional practices are important for reducing greenhouse gas emissions, for example through energy efficiency, support for renewables, and changes to urban form. Yet other spaces and modes of deliberation are needed to address how climate impacts, energy use, and climate justice are tied to more fundamental economic, social, and cultural structures; and the worry is that where these more fundamental structures go unaddressed, mainstream responses to climate change may treat symptoms rather than causes (Szeman and the Petrocultures Group 2016). Tensions between reforming existing systems and advocating for broader structural change are not easily resolved, and specific mini-publics should not be made to carry these burdens alone. This is a tension that informs not only debates about public deliberation but also about climate change more generally, and the tension merits a diversity of practical and pedagogical responses.

Conclusion

Public deliberation is debate and discussion aimed at producing reasonable, well-informed recommendations and plans for action, rooted in the values of participants. That simple description masks disagreement in the field, however, including tensions between bottom-up and top-down approaches to deliberation; how much influence deliberation needs to have in order to be legitimate; the fit between instances of deliberation and building deliberative societies; the places of emotion, narrative, and rhetoric in public deliberation; questions of framing; relationships between deliberation and activism; and how the professionalization of the field of public deliberation influences what happens on the ground. ABCD's work tangled with all these disagreements.

There also is a tangled relationship between public deliberation and the challenges of climate change. Climate change can be framed in multiple ways, none of them innocent of power considerations: this points to the importance of inclusive, collaborative approaches to the issue, but also to why such approaches are so fraught. Diversity is not just about demographic representativeness: rather, it is about looking at relations of inequality, oppression, and marginalization between social groups. Perspectives on climate change are shaped by such memberships, and we have suggested that the design of climate deliberations needs to address power within and between social groups. The framing of climate deliberations also risks reiterating dominant narratives, and this may be aggravated when deliberations are oriented toward policy influence and political decision makers.

Public deliberation can lead to a "disturbance of everyday reasoning habits" as people are "jolted out" of the routine scripts that organize their lives (Ryfe 2005, 56–57). We have highlighted some of the political choices that enable deliberations on climate change to disturb everyday reasoning about climate change, or to reiterate it. Other chapters in this volume offer further reflections on the politics of climate change deliberation, the relationships of politics to process design, and how ABCD navigated these politics within multiple constraints.

Note

1. The authors of this section are academics who have connected deliberative democratic theory to practical projects. Gwendolyn Blue is a cultural geographer with research interest in public engagement with science and technology. She was a researcher and site organizer of World Wide Views on Global Warming, the first global scale public engagement initiative on climate change, and was academic lead on one of ABCD's deliberations, Water in a Changing Climate. David Kahane is a political theorist specializing in deliberative democracy who spent seven years as a collaborator on a project studying the effectiveness of citizen participation as a means to pro-poor political outcomes in the global south (http://www.drc-citizenship.org). He was also the Project Director of Alberta Climate Dialogue, and one of the key designers of one of ABCD's deliberations, the Citizens' Panel on Edmonton's Energy and Climate Challenges.

References

Adkin, Laurie, Lorelei Hanson, David Kahane, John Parkins, and Steve Patten. 2016. "Can Public Engagement Democratize Environmental Policy-Making in a Resource-Dependent State? Comparative Case Studies from Alberta, Canada." *Environmental Politics*. doi:10.1080/09644016.2016.1244967.

Blue, Gwendolyn. 2015. "Framing Climate Change for Public Deliberation: What Role for the Interpretive Social Sciences and Humanities?" *Journal of Environmental Policy and Planning* 18(1): 67–84.

Blue, Gwendolyn, and Jacquie Dale. 2016. "Framing and Power in Public Deliberation with Climate Change: Critical Reflections on the Role of Deliberative Practitioners." *Journal of Public Deliberation* 12(1): art. 2. http://www.publicdeliberation.net/jpd/vol12/iss1/art2.

Blue, Gwendolyn, Jennifer Medlock, and Edna Einseidel. 2012. "Representativeness and the Politics of Inclusion." In *Citizen Participation in Global Environmental Governance*, edited by Richard Worthington, Mikko Rask, and Lammi Minna, 139–52. London: Earthscan.

Bohman, James. 1996. *Public Deliberation*. Cambridge: MIT Press.

Chilvers, Jason, and Noel Longhurst. 2012. *Participation, Politics, and Actor Dynamics in Low Carbon Energy Transitions*. Norwich, UK: University of East Anglia Science Society and Sustainability.

Cornwall, Andrea, and John Gaventa. 2001. "From Users and Choosers to Makers and Shapers: Repositioning Participation in Social Policy." *IDS Working Paper 127*. Brighton, UK: Institute of Development Studies.

Dryzek, John. 2006. *Deliberative Global Politics: Discourse and Democracy in a Divided World*. Cambridge, MA: Polity Press.

———. 2013. *The Politics of the Earth: Environmental Discourses*. 3rd ed. Oxford: Oxford University Press.

Fishkin, James. 1997. *The Voice of the People: Public Opinion and Democracy*. New Haven, CT: Yale University Press.

Fung, Archon. 2003. "Survey Article: Recipes for Public Spheres: Eight Institutional Design Choices and Their Consequences." *Journal of Political Philosophy* 11: 338–67.

———. 2005. "Deliberation Before the Revolution: Towards an Ethics of Deliberative Democracy in an Unjust World." *Political Theory* 33(2): 397–419.

Gaventa, John. 2006. "Finding the Spaces for Change: A Power Analysis." *IDS Bulletin* 37: 6.

Gaventa, John, and Gregory Barrett. 2010. "So What Difference Does It Make? Mapping the Outcomes of Citizen Engagement." *IDS Working Paper 347*. http://www.ids.ac.uk/files/dmfile/Wp347.pdf.

Habermas, Jürgen. 1993. *Justification and Application: Remarks on Discourse Ethics*. Cambridge: MIT Press.

Hendriks, Carolyn, and Lyn Carson. 2008. "Can the Market Help the Forum? Negotiating the Commercialization of Deliberative Democracy." *Policy Sciences* 41: 293–313.

Hulme, Michael. 2009. *Why We Disagree About Climate Change*. Cambridge: Cambridge University Press.

Kahane, David. 2002. "Délibération Démocratique et Ontologie Sociale." *La Démocratie Délibérative*, Numéro spécial de *Philosophiques* 29(2): 251–86.

Kahane, David, Kristjana Loptson, Jade Herriman, and Max Hardy. 2013. "Stakeholder and Citizen Roles in Public Deliberation." *Journal of Public Deliberation* 9(2): art. 2. http://www.publicdeliberation.net/jpd/vol9/iss2/art2.

Kettering Foundation. 2011. "Naming and Framing Difficult Issues to Make Sound Decisions." A Kettering Foundation Report. https://www.kettering.org/sites/default/files/product-downloads/Naming_Framing_2011-.pdf.

Lee, Caroline W., Michael McQuarrie, and Edward T. Walker. 2015. *Democratizing Inequalities: Dilemmas of the New Public Participation*. New York: New York University Press.

Lee, Carolyn. 2011. "Five Assumptions Academics Make about Public Deliberation, and Why They Deserve Rethinking." *Journal of Public Deliberation* 7(1): art. 7. http://www.publicdeliberation.net/jpd/vol7/iss1/art7.

———. 2015. *Do-It-Yourself Democracy: The Rise of the Public Engagement Industry.* Oxford: Oxford University Press.

Leighninger, Matt. 2010. *Creating Spaces for Change: Working Toward a 'Story of Now' in Public Engagement.* Battle Creek, MI: Kellogg Foundation.

Lukensmeyer, Carolyn. 2012. *Bringing Citizen Voices to the Table: A Guide for Public Managers.* San Francisco: Jossey-Bass.

Mansbridge, Jane, James Bohman, Simone Chambers, Thomas Christiano, Archon Fung, John Parkinson, Dennis F. Thompson, and Mark E. Warren. 2012. "A Systemic Approach to Deliberative Democracy." In *Deliberative Systems: Deliberative Democracy at the Large Scale*, edited by John Parkinson and Jane Mansbridge, 1–26. Cambridge: Cambridge University Press.

Moser, Susanne, and Dilling, Lisa. 2010. "Communicating Climate Change: Closing the Science – Action Gap." In *The Oxford Handbook of Climate Change and Society*, edited by John Dryzek, R.B. Norgaard, and D. Schlosberg, 161–74. New York: Oxford University Press.

Nabatchi, Tina, and Matt Leighninger. 2015. *Public Participation for the 21st Century.* San Francisco: Jossey-Bass.

Oreskes, Naomi, and Erik M. Conway. 2010. *Merchants of Doubt.* London: Bloomsbury Press.

Pallett, Helen, and Jason Chilvers. 2013. "A Decade of Learning about Publics, Participation, and Climate Change: Institutionalising Reflexivity?" *Environment and Planning* A 45(5): 1162–83.

Ryfe, David. 2005. "Does Deliberative Democracy Work?" *Annual Review of Political Science* 8: 49–71.

Szeman, Imre, and the Petrocultures Group. 2016. *After Oil.* Morgantown: West Virginia University Press.

Ulrich, Werner. 2005. "A Mini-Primer of Boundary Critique." http://wulrich.com/boundary_critique.html.

Williams, Melissa. 1998. *Voice, Trust, and Memory: Marginalized Groups and the Failings of Liberal Representation.* Princeton, NJ: Princeton University Press.

Young, Iris Marion. 2005. *Inclusion and Democracy.* Oxford: Oxford University Press.

The Economic and Political Context of Climate Policy in Alberta

Geoff Salomons and John R. Parkins

This chapter situates the Alberta Climate Dialogue (ABCD) deliberations within the political and economic context of the province of Alberta. We argue that overall the Alberta context is one that is generally resistant to public participation mechanisms. When public engagement is undertaken it is often designed to secure public acceptance of policy proposals rather than meaningful input into the design of such policies. We also note in this chapter a tension between high-profile provincial deliberations and low-profile localized deliberations that are less risky for conveners but also potentially less effective in forging policy alternatives. Despite these general tendencies, there are exceptions where political leaders and civil servants are genuinely open to more innovative approaches to public engagement. Some of the work by ABCD reflects these positive outcomes. By outlining the contextual challenges ABCD faced, it is our hope that other organizations seeking to design deliberative processes will gain a better understanding of how history and context inform the design and implementation public engagement.[1]

The overview presented is this chapter uses a multi-scalar geopolitical approach to the political and economic context within which ABCD operated. The four citizen deliberations that ABCD members participated in were aimed at different levels of governance: municipal, regional, and provincial. We delineate the geopolitical factors at different scales and discuss key drivers and trends, recognizing the interconnections among the levels and how they shaped the operations of ABCD.

Political and Economic Context at Multiple Scales

International

Attempts to forge a global climate policy, primarily through the United Nations Framework Convention on Climate Change, have met with limited success (see the introduction). While agreements were initially formed and the creation of the Kyoto Protocol was thought to be at least a modest first step, the legacy of these multinational initiatives has been little more than agreements based on the lowest common denominator. As Harrison and Sundstrom (2010) argue, the primary reason for limited progress is that many nations came to the meetings (in establishing the Kyoto Protocol in particular) without completing the domestic political legwork. The legacy of Kyoto has thus been mixed, with some jurisdictions achieving their emission reductions (e.g., the European Union), some ratifying the protocol but failing to take action to achieve the agreed-upon reductions (e.g., Canada) and some failing to gain domestic ratification (e.g., United States).

Attempts to develop an agreement to replace the Kyoto Protocol floundered in Copenhagen in 2008. The Canadian government's approach to the international negotiation process under Conservative Prime Minister Stephen Harper could be characterized as minimalist at best and obstructionist at worst. While Canada is not a major player on the international scene, and essentially acquiesced in climate policy decisions taken by the United States, it also used the climate negotiations forum as an opportunity to defend the continued development and expansion of Alberta's oil sands. In 2011, owing to a decade of failure by both Liberal and Conservative governments to enact any meaningful policies aimed at reducing emissions, Canada formally withdrew from the Kyoto Protocol (Curry and McCarthy 2011).

Federal

At the federal level in Canada, it was the Liberal government under Prime Minister Jean Chrétien that signed and ratified the Kyoto Protocol in 2002. However, very little was done in the following four years to achieve the results to which Canada had agreed. In 2006, a reunited Conservative Party of Canada, with strong support from its base in Alberta, formed a minority government. The party was staunchly pro-oil sands development. In the 2008 election, climate change was one of the central campaign issues, with all three major federal parties—Liberal, Conservative, and New Democratic—campaigning for some form of carbon pricing. The Liberals, under party leader Stéphane Dion, campaigned for a carbon tax as part of a portfolio of climate-friendly policies called "Green Shift." The Conservatives and New Democratic Party proposed a cap-and-trade system. In the 2008 election, the Conservative Party secured another minority government, and in the same year, the global recession eclipsed climate change as an issue of concern as governments around the world scrambled to address more immediate economic concerns. In the 2011 election, the Conservative Party finally secured a majority government and further stalling on progressive climate policies ensued.

Despite repeated promises, the federal government continually delayed the implementation of greenhouse gas (GHG) emission reduction policies. Rather than an economy-wide price on carbon, the Canadian government under the Conservatives adopted a "sector-by-sector" regulatory approach to regulating GHG emissions (Government of Canada 2015). Prior to the federal election in October 2015, the Canadian government developed regulations on light and heavy transportation and on coal-generated electricity. According to the Canadian government, in 2011 these regulations covered approximately 30 per cent of Canada's GHG emissions (Enviroment Canada 2014).

In addition to weak regulations on GHG emissions, Prime Minister Harper's government was also hostile to environmental concerns. The day before the Joint Review Panel hearing opened on the proposed Northern Gateway pipeline, which would go from Alberta to the British Columbia coastline, Natural Resource Minister Joe Oliver (2012) issued an open letter calling anyone who opposed oil and gas infrastructure "radicals" with an "ideological agenda." The ensuing 2012 budget included changes to Canada's Environmental Assessment Act, which limited public participation and streamlined the process to allow for more timely approvals of major industrial or natural resource projects (Salomons and Hoberg 2014). These changes were later discovered to have been at the request of the oil industry (Paris 2013). The 2012 budget also included additional funding for the Canada Revenue Agency to conduct audits of charities to ensure compliance with legal limits of political activity such as lobbying. In Canada, charitable organizations are not allowed to spend more than 10 per cent of their budget on political activity. To date, the only charities that have been targeted for audits are those highly critical of the Conservative administration (Solomon and Everson 2014).

As project activities related to ABCD were winding down, the 2015 federal election ushered in a major shift in federal climate policy, with a majority government for the Liberal Party. Prime Minister Justin Trudeau quickly moved to meet with provincial premiers to discuss federal climate policy in time for the twenty-first Conference of the Parties meeting (COP 21) in Paris. Prime Minister Trudeau hosted a first ministers' meeting with the Canadian provincial premiers on March 3, 2016, to discuss how Canada will meet its international climate change obligations (Prime Minister of Canada's Office 2016). The discussions focused primarily on ways to price carbon in Canada.

Provincial

Under the long-standing leadership of the Progressive Conservative Association of Alberta, the political and economic context of climate change policy during ABCD's operation was just as regressive as the federal one. Indeed, part of the interest in locating the ABCD research project in Alberta was the difficult context, with polarized opinions, entrenched interests, and government resistance to progressive climate change policies. Could deliberative democratic approaches to policy making move the dial on climate policy within the province? If so, Alberta could thus serve as a crucial example of applying deliberative processes to address intractable political issues. If deliberation were deemed successful in Alberta, the prospects for success of citizen deliberations on climate change policy development elsewhere could be encouraged and bolstered (Seawright and Gerring 2008).

The central economic issue in Alberta is oil and gas development. With the advent of in situ technology[2] that can extract significant amounts of oil sands resources without the need for strip mining, Alberta's proven reserves ballooned from just under 5 billion barrels in 2002 to 180 billion barrels in 2003, according to the International Energy Agency's estimates. With these reserves, Canada ranks third in the world behind Saudi Arabia and Venezuela, so the economic potential of the oil sands is vast. Even with oil prices at forty dollars per barrel, the total potential value of the oil sands is C$7.2 trillion, almost four times the size of the Canadian economy in 2014 (Statistics Canada 2016). This economic value presents a challenge for Canada, as it offers great potential to sustain wealth creation and stable revenues for the government, but at high environmental costs. Moreover, scholars consistently highlight that economic interests often hold a privileged position within democratic political process (Lindblom 1977). Political theorist John Dryzek (1995, 15) succinctly sums up the problem:

> All liberal democracies currently operate in the context of a capitalist market system. Any state operating in the context of such a system is greatly constrained in terms of the kinds of policies it can pursue. Policies that damage business profitability—or are even perceived as likely to damage that profitability—are automatically punished by the recoil of the market. Disinvestment here means economic downturn. And such downturn is bad for governments because it both reduces the tax revenue for the schemes those governments want to pursue (such as environmental restoration), and reduces the popularity of the government in the eyes of the voters. This effect is not a matter of conspiracy or direct corporate influence on government: it happens automatically, irrespective of anyone's intentions.

In oil and gas resource-abundant jurisdictions, there is the possibility for those economic interests to be magnified and the government "to exhibit the economic and political characteristics of a petro-state" (Homer-Dixon 2013). While various definitions exist, a petro-state is often defined either economically (an oil-producing jurisdiction with typically more than 30 per cent of revenue coming from oil revenue) or politically, with behaviour favourable to the interests of oil companies (Karl 1997; Ross 2012; Mitchell 2011). Thus, while economic data at the national level might lead some to question whether *Canada* is properly labelled a petro-state (Leach 2013), at the provincial level, where jurisdiction over natural resources lies, we see a different story. Since 1971, oil revenue as a percentage of total provincial government revenue has varied significantly but over the past decade has hovered between 20 to 30 per cent of total revenue (see Figure 3.1). The energy sector currently comprises approximately 23 per cent of Alberta's GDP and over 75 per cent of its exports (Government of Alberta 2015b). Alberta also implemented a very low tax regime that features the lowest corporate tax rates in Canada (Government of Alberta 2015a) and, until 2015, had a flat personal income tax rate of 10 per cent. Alberta is the only jurisdiction in Canada without a provincial sales tax (Government of Alberta 2016). Despite these low revenue streams, program spending on health, education, and other services has typically remained comparable to that of other provinces (Taft et al. 2012).

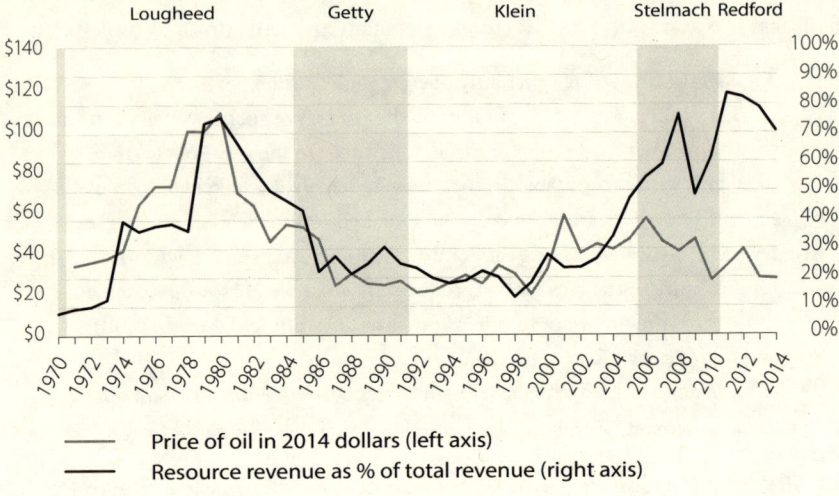

Figure 3.1. Resource revenues as a percentage of total revenue from 1970 to 2013 in Alberta, with oil prices in 2014 dollars.

Source: Resource revenue data from Alberta Energy Resource Revenue Workbook; Oil price data from BP Statistical World Energy Handbook.

In highlighting the "petro-state" label, we are less concerned with whether Alberta is rightly placed in the same category as other "petro-states" such as Venezuela or Saudi Arabia. As Shrivastava and Stefanick (2015, 12) suggest, "by explaining development performance solely in terms of the size and nature of the resource wealth, the oil and democracy literature often does not adequately account for the role of internal and external social, political, and economic environments in shaping development outcomes in resource-abundant countries." Within this context our goal is to emphasize the privileged place that the oil and gas industry has within the Alberta political-economic context that distinguishes it from other jurisdictions. This privileging of the oil and gas industry has several ramifications germane to ABCD's work.

According to leading petro-state scholar Terry Lynn Karl (1997), one of the political ramifications of petro-state politics is regime stability. While often referring to non-democratic regimes that are able to use resource revenue to appease or suppress opposition (e.g., members of OPEC), this stability can also emerge within democratic jurisdictions. Resource revenue augments budget revenue, allowing democratic governments to offer increased program spending while keeping taxes low. This short-sighted approach appears to reflect sound fiscal policy when revenues are high but creates significant budgetary deficits when commodity prices are low (Ryan 2013).

One of the consequences of such regime stability with regard to public deliberations broadly speaking is that public engagement processes can be employed to provide a democratic façade over decisions already made by the administration. Without significant electoral competition, there is little impetus to genuinely seek out and identify citizen preferences. Instead, public engagement offers a means of selling projects to citizens or facilitating the identification of desired outcomes (Davidson and MacKendrick 2004).

A second ramification of this petro-state behaviour is that it allows accountability linkages between the government and its citizens to be supplanted by accountability linkages between the government and the oil industry. The government becomes less responsive to the preferences of its citizens and more responsive to the preferences of oil companies (Karl 1997).

Third, the privileged place and the power of the oil industry in this context negatively impacts the democratic quality of governance in Alberta, a negative effect noted by a number of scholars (Shrivastava and Stefanick 2012, 2015; Adkin 2016). This democratic decline creates additional points of resistance to democratic innovations, particularly ones that are intentionally aimed at progressive climate policy development.

Finally, while not causally related to "petro-states," Alberta has a unique political culture, predating the ascent of the oil industry. Jared Wesley (2011) argues that an overarching political code or culture based on autonomy, populism, and individualism exists within Alberta, and that those politicians and parties that have been most successful are those that tap into and actively cultivate this code. Electorally speaking, Alberta has been a conservative fortress, as the Progressive Conservatives held power for forty-four years. While in 2015 the left-leaning New Democratic Party swept into power, it is too early to tell whether this shift indicates a longer-term change in Alberta's political culture (whether due to changing values or changing demographics) or is a one-off protest vote against the reigning Progressive Conservatives. At the federal level, electoral support in Alberta has consistently leaned to the right. No other province in Canada has had such a stable preference for conservative-leaning politics.

As expected, this political culture does not lend itself to significant action on climate change, especially if such action would potentially threaten the oil sands as the economic engine of the province. Polling data suggests that Albertans are more likely than other Canadians to be skeptical of climate change (Forum Research 2014), believe too much attention is paid by the federal government to climate change (Angus Reid Institute 2015), believe Canada is doing more than the rest of the world to deal with climate change, and have the most opposition to a carbon tax (Environics Institute for Survey Research and David Suzuki Foundation 2014).

Municipal

Municipalities face a wide variety of barriers to action on climate change, including more immediate priorities, lack of information, lack of capacity, and lack of knowledge (Robinson and Gore 2005). Moreover, at the municipal level, there is no single policy instrument that can leverage emission reductions over the whole of a municipality's jurisdiction. Rather, municipalities must take the problem of climate change and use it as a lens for various aspects/areas of municipal policy making, areas which are also subject to other lenses, problem definitions, and other forms of contestation (City of Edmonton 2011). Despite these barriers, action at the municipal level has the potential to significantly impact climate change mitigation for a number of reasons.

First, as urbanization trends continue, the United Nations (UN) predicts that by 2030 approximately 60 per cent of the global population will reside within urban areas. The UN's Intergovernmental Panel on Climate Change's (IPCC) *Fifth Assessment Report* notes that urban areas account for approximately 67 per cent to 76 per cent of global energy use (IPCC 2014). Urbanization leads to an increased emissions footprint for various reasons, whether it is increased consumption dependence on certain forms of transportation as a consequence of urban form or increased transportation required for goods (Satterthwaite 2009).

Second, municipalities have significant influence over the GHG emissions within their jurisdictions. A report drafted for the Federation of Canadian Municipalities estimates that in 2006 Canadian municipalities had direct or indirect control over approximately 44 per cent of national greenhouse gas emissions in Canada (EnviroEconomics 2009). Policy areas such as urban form (sprawl versus densification), transportation (i.e., roads, transit, cycling infrastructure), building codes, waste management, and commercial and industrial development all have an impact on GHG emissions. While the complexity of coordinating all these policy areas to address a single issue such as climate change is daunting, it is the municipalities which have jurisdiction, and so it is up to them to act on these areas.

Third, the immediacy of the municipal context provides people with more tangible and concrete projects with which to work. This immediacy and concreteness has the potential to overcome the various psychological impediments to climate action (e.g., Gifford 2011; Rachlinski 2000; Weber 2011). When unprecedented flooding along the Bow River in 2013 caused massive damage in Calgary, High River, and a number of other southern Alberta communities, climate change adaptation became a much less abstract issue for many municipalities.

Finally, knowledge and capacity barriers at the municipal level can potentially be addressed through coordination and collaboration between municipalities. International associations such as International Council for Local Environmental Initiatives (ICLEI) have worked to help municipal governments coordinate, collaborate on, and disseminate strategies for environmentally sustainable policies and programs (ICLEI 2017). In Canada the Federation of Canadian Municipalities (FCM) also has a Green Municipal Fund, which allows for funding and training for sustainability initiatives at the local level, thus increasing the capacity of cities to implement policies and programs for climate mitigation and adaptation (FCM 2015). This capacity at the municipal level allows for a streamlining of climate change mitigation and adaptation policies, multiplying their effectiveness and reducing the costs of implementation.

Consulting on Environmental Regulation in Alberta

Public deliberation on environmental policy in Alberta coincides with several waves of international environmentalism as well as several waves of intensive resource development over the last fifty years. In the late 1960s and early 1970s, the Province of Alberta enjoyed a wave of populist environmentalism that translated into several remarkable provincial initiatives. Among them was the formation of the Environmental Conservation Authority in 1972 with broad powers and a public advisory committee mechanism to respond to growing concerns by residents across the province regarding water quality issues and oil industry development. This new body made extensive use of multiple public advisory committees that were "semi-independent bodies comprised of volunteers who chose to study any issue they deemed important, passing resolutions accordingly" (Stefanick and Wells 2000, 370). Coupled with these early regulatory innovations, the research and practitioner community hosted an international conference on participatory approaches to environmental decision making in 1977 which resulted in a two-volume proceedings containing fifty papers on the emerging theoretical and practical aspects of public involvement in the development of environmental policy and regulation. In these proceedings the author notes that "a diverse array of formal channels are now open to individual citizens who wish to become involved in matters of public policy" (Sadler 1977, 2).

Although this early enthusiasm for an open, inclusive, democratic approach to environmental policy development was met with subsequent pushback and retrenchment from government and industry at various junctures during the 1980s and 1990s, a key point in this history involves persistent tensions between the centralized, command-and-control aspect of government regulation at the provincial scale and the decentralized and market-oriented approaches to environmental governance at regional and local scales. Examples of public consultation at the provincial scale include the development of the Alberta Forest Conservation Strategy in 1995 and Special Places 2000 (Schneider 2001; Stefanick and Wells 2000). Contrasting these approaches are local initiatives such as community-based public advisory committees in the forest sector (Parkins 2006) and similar processes in the energy sector. For example, regional approaches to public engagement are reflected in the "synergy groups." Synergy Alberta has a mission to foster and support "mutually satisfactory outcomes in Alberta communities by providing information, mutual learning, communication, skill development, facilitation and resources" (Synergy Alberta 2016).

Regional public consultation is also linked to land-use planning processes in Alberta (Parkins 2011). The Land-use Framework was officially launched in December 2008 with seven specific planning regions. According to the Alberta Land Stewardship Act (Government of Alberta 2009, 5), land-use planning is intended "to create legislation and policy that enable sustainable development by taking account of and responding to the cumulative effect of human endeavor and other events." The heart of this planning process includes a Regional Advisory Council, a multi-stakeholder group intended to bring forward local insights and perspectives on land-use issues. Other regional governance mechanisms include the province's Water for Life Strategy—released in 2003—which provides the opportunity for regional stewardship organizations to give feedback to the province regarding the governance of water resources (Government of Alberta

2003, 2008). One such organization based in southern Alberta, the Oldman Watershed Council, partnered with ABCD to host Water in a Changing Climate, a one-day deliberation on water governance and climate change adaptation.

One of the great strengths of regional initiatives such as the Land-use Framework or the Oldman Watershed Council is that people who are most directly impacted by land- and resource-use policies (who have material interests) have a say in the process and can contribute a sense of local knowledge and local values to improve decision making. There is also good reason to enact environmental policy based on ecological boundaries as opposed to more arbitrary geopolitical boundaries. Yet, as scholars have noted, the problem here is that "much of what passes under the rubric of stakeholder involvement has more to do with assuring and legitimating the goals of sponsoring managers than introducing new perspectives and knowledge or empowering those who occupy the spectator mainstream or live on the margins of community and society" (Kasperson 2006, 321). In other words, local stakeholder processes are vulnerable to local political and economic elites, particularly when publics in these regions are directly dependent on resource industries for their livelihood (Parkins and Sinclair 2014).

Consulting on Climate Policy in Alberta

Reflecting a centralized province-wide approach to public engagement on environmental issues, climate policy in Alberta is summarized here with reference to Adkin (2014). Her paper offers important context for the work of ABCD. One key aspect of policy development on climate change is the international flavour of public concern and political pressure. From international Indigenous rights organizations making note of changing impacts on Indigenous cultures in the Arctic, to growing levels of concern expressed by climate scientists, Alberta's lack of response to climate challenges has been noted outside the province and the country (Adkin 2014).

National and international pressure resulted in two large-scale public consultations in Alberta, one in 2002 and a second in 2007. In response to federal government GHG emission reduction targets as agreed to in the Kyoto Protocol, in 2002 the Province of Alberta opted for a policy that established emissions intensity targets. Stakeholder consultations at that time "served mainly to make sure that the draft policy was acceptable to representatives of large emitters, and to learn what kinds of objections could be expected from ENGOS" (Adkin 2014, 6).

Beyond the involvement of key stakeholders, consultation was limited in this 2002 consultation to an online survey with approximately 260 participants. At the same time, the provincial government worked hard to convince Albertans that implementation of the Kyoto Protocol, and the implementation of policies necessary to achieve reductions, would mean the destruction of the Alberta economy. Summarizing the efforts of politicians, a government minister notes, "Alberta was vocal in its opposition to ratifying the Protocol and undertook an initiative to call for an alternate solution to climate change that was 'made in Canada'" (Government of Alberta 2003, 4).

Following the release of the IPCC's *Fourth Assessment Report*, the context for public consultations in 2007 was somewhat different from the process five years earlier. The IPCC report reiterated the dire consequences for inaction on climate change, and the salience of climate change concern was palpable. Al Gore's movie *An Inconvenient Truth* created a broader public awareness of the climate challenge, and all the federal parties in Canada proposed serious carbon pricing policies in the run-up to the 2008 national election, as noted above (Harrison 2012). In 2007, the extent of broad-scale public engagement was more significant, with 2,600 responses to online workbooks, but the focused remained squarely on the interests of key stakeholders. Two multi-stakeholder roundtables were held, with emphasis on economic interests in the province (Adkin 2014). The consultation did show, however, a significant shift in public interest and perception of climate change wherein "a strong majority of respondents expected government action on this issue to include absolute emission targets" (McMillan 2007, 1, as quoted in Adkin 2014). In summarizing the legacy of these public consultations on climate policy in Alberta, Adkin concludes that "the impetus for climate change policy has not come from provincial political leadership, the existence of a strong left or green party, the importance of agriculture or ecotourism to the province's economy, awareness of the foreseeable costs of global warming for the provincial economy, the efforts of a handful of publicly engaged scientists, or the small but persistent environmental community" (Adkin 2014, 19). Climate change policy development in Alberta occurred in response to international and national pressures, resulting in centralized and stakeholder-based public consultations that were managed in order to achieve particular outcomes that would not negatively impact the energy sector.

A key regulatory change in Alberta during this time was the Specified Gas Emitters Regulation (SGER), which came into effect in 2007. This regulation requires any large facility that emits over 100,000 tonnes of GHGs to reduce its emission intensity. Failure to achieve such reductions required the company to either purchase offsets or pay fifteen dollars per tonne into a Climate Change and Emissions Management Fund for emissions that exceed the facility's emission reduction target. This regulation essentially put a price on carbon for large emitters. However, restricting the policy to a handful of large facilities and the minimal charge per tonne levied for GHG emissions limit the efficacy of this policy (Dyer et al. 2011). Effective GHG emissions reductions would require higher carbon pricing, more in line with British Columbia's economy-wide carbon tax of thirty dollars per tonne.

According to the Alberta government's own emissions modelling at the time, it hoped to achieve its greatest emissions reductions through investment in carbon capture and storage (CCS) technology (Dyer et al. 2011, Figure 1). CCS technology sequesters carbon emissions from large GHG-emitting facilities, such as coal-fired electricity generation facilities, and stores that carbon underground, thus removing the emissions from the atmosphere. The Alberta government under the Progressive Conservatives allocated 1.5 billion dollars to fund two large-scale CCS projects. The hope was that demonstrating their technical feasibility would encourage their adoption by other companies, but without adequate financial incentives (such as a substantial price on carbon) other CCS projects were not financially viable, and in 2014 the program was scrapped.

Implications for ABCD

Reflecting on this context for public consultations on climate policy in Alberta, there are several notable implications for the evolution of public engagement processes within the ABCD project. Given the launch of this project in 2010, it is noteworthy that the previous ten years were marked by a shift toward the right in federal politics and ongoing intransigence and delays in formulating federal leadership on climate policy. Coupled with this national context, the Government of Alberta extended their foot dragging with high-profile campaigns against ratification of the Kyoto Protocol, several rounds of province-wide consultations with stakeholders, and limited progress on meaningful policy development addressing climate challenges. All of this took place amid growing recognition worldwide of the scientific evidence and the obvious impacts of climate change, not to mention international awareness of Alberta's role as a carbon-intensive energy producer.

In some respects, the intransigence of Albertans is not surprising. A stable and conservative political culture, along with recognition among many Albertans that strong climate policy would negatively impact the energy sector and their way of life, caused understandable concern about the implications of strong climate policy for the future of the province. Yet, by the end of the 2000s, surveys showed that Albertans were growing more aware of the need for strong climate policy. Going into the year 2010, at the inception of the ABCD project, there was palpable concern within policy communities about the lack of progress on climate policy in Alberta, coupled with optimism and a sense that perhaps Alberta was ready to take on more meaningful and more vigorous climate change responses.

One of the initial objectives within ABCD was to work toward implementing a high-profile provincial dialogue on climate change. In the end, ABCD was involved in and sponsored a series of municipal, regional, and topical deliberations. There are several reasons for this shift in approach within the project, but one aspect relates to the shift in public consultation more generally within the province. In the 1990s, the Alberta government initiated a number of province-wide consultations on topics ranging from forest policy to protected areas and climate change, but these high-profile consultations were often heavily criticized and did not always yield the kind of social acceptance that government and industry leaders were hoping to achieve (e.g., Stefanick and Wells 2000). In response, the provincial government moved toward regional approaches to public consultation on environmental issues (Parkins 2006). In line with this shift, the ABCD project found success in adopting a localized and focused approach to public engagement.

Looking more specifically at the four citizen deliberations ABCD members participated in, we see a number of ways in which the context shaped those events. Recognizing the unique position that municipalities have in relation to climate policy, ABCD looked for opportunities to partner with the City of Edmonton. The first deliberation that ABCD members participated in was Edmonton's City-Wide Food and Agriculture Citizen Panel (Food and Ag Panel). This deliberation was convened by the Centre for Public Involvement (CPI), a not-for-profit organization jointly funded by the University of Alberta and the City of Edmonton that provides leadership on effective methods of public involvement, mostly for city initiatives. The purpose of the Food and Ag Panel was to provide input into the development of an initiative called *fresh*, Edmonton's food and urban agriculture strategy (Centre for Public Involvement 2012).

The inception of *fresh* addressed two quite different spatial factors, one a global trend and the other a local development. First, *fresh* had its roots in the growing interest in urban agriculture and food security internationally and nationally that has increasingly been finding tangible policy expression as an urban planning issue (Kaufman 2009). Second, *fresh* came about through the efforts of a local not-for-profit network, the Greater Edmonton Alliance (GEA). GEA built public awareness and support for expanded food security and the preservation of prime farmland within Edmonton's municipal boundaries, and very successfully mobilized hundreds of Edmonton residents to attend a series of public meetings to express these desires to Edmonton's City Council. Initially Edmonton's mayor and council offered their support, if not enthusiasm, for the development of a food and urban agriculture strategy that would include extensive public involvement. However, as it became apparent that the majority of the public saw a food strategy as the mechanism to protect Edmonton's remaining farmland from development, many members of Council and the mayor at the time (a Conservative and a real estate developer) became less enthusiastic about the food strategy and the associated public participation processes. The farmland in question has some of the best agricultural soils in Canada (Hanson and Schrader 2016), but the area was also identified as an ideal residential and service location to support adjacent chemical refining and manufacturing industries (KlineGroup 2008). Accordingly, the area was approved for development in 2010.

The citizen deliberation associated with *fresh* was a great success on many fronts, but it had little impact on the final food and urban agricultural policy process. The recommendation with the greatest support from the Food and Ag Panel was to "create and/or amend zoning, bylaws, fees, and taxes to prohibit developments on good fertile agricultural land, particularly the northeast farmland" (City of Edmonton 2012, 2). While this recommendation echoed the results of other public engagement processes associated with the development of *fresh*, the final draft of the food and urban agriculture strategy, written by City of Edmonton staff, did not mirror this emphasis. Instead, following the mayor's very vocal lead, City Council approved a vague strategy that included no protection for agricultural lands within Edmonton's boundaries (Hanson and Schrader 2016). The result of this public process was disillusioning for some, and disappointing for many others. It reinforced a sense that the government remains largely beholden to petro-state politics whereby the interests of oil and gas are placed ahead of environmental, social, and other resource management concerns (Waller 2012).

While some members of the ABCD team undertook research on Edmonton's Food and Agricultural Strategy, more intensive efforts were focused on ABCD partnering with the City of Edmonton's Office of the Environment on a long-term policy response to city energy and climate challenges. The high-level environmental strategic plan, *The Way We Green*, was the overarching framework within which the City of Edmonton developed a climate action strategy (City of Edmonton 2011). The Citizens' Panel on Edmonton's Energy and Climate Challenges, developed jointly by ABCD and the City of Edmonton, involved citizen-based discussions on five policy levers that the Office of the Environment determined had the greatest potential for reducing Edmonton's greenhouse gas emissions and dependence on fossil fuels (Pembina Institute and HB Lanarc 2012).

The Citizens' Panel on Edmonton's Climate and Energy Challenges (Edmonton Panel) was ABCD's largest and most extensive deliberation. City administrators viewed the Edmonton Panel as an opportunity to build public support, and thereby Council support, for the strategy. In an interview with an ABCD member, city administrator and lead author of the Energy Transition Strategy, Jim Andrais (2015) stated, "I think that without [the Edmonton Panel] we don't even get into City Council." The Edmonton Panel made key contributions to the Energy Transition Strategy, including adopting guiding values that were reflected in the city's strategy, and perhaps more importantly, instilling confidence in city administrators and Council that they would have the support of Edmontonians in approving and implementing this broad strategy that would impact various aspects of city life. As noted by Robinson and Gore (2005), muncipalities face wide-ranging challenges in developing climate policy; the Edmonton Panel overcame public knowledge and information barriers by linking energy and climate challenges to issues such as housing, transportation, and industrial development.

The relative success of the Edmonton Panel, including the fact that it was able to overcome some of the contextual factors operating against such an initiative that have been outlined above, is the result of several factors. First, ABCD was able to partner with a jurisdiction already committed to pursuing more robust climate policies. ABCD was able to link up with an already existing policy process and show how public engagement could strengthen, encourage, and embolden elected officials to take strong action. While some jurisdictional questions emerged, such as how the City could encourage the greening of the provincial electricity grid, the central policy levers under consideration were well within the City's control.

The second factor was electoral results that reinforced the climate policy agenda. At the beginning of the project, ABCD's main liaison with City Council was then Councillor Don Iveson, who was tasked with taking up the city's environment and sustainability initiatives. After the Edmonton Panel, a municipal election was held and Don Iveson was elected mayor of Edmonton. This fortuitous event placed an individual at the centre of municipal decision making who was in the room during the Edmonton Panel and understood the quality of deliberation that occurred. It allowed Mayor Iveson to be a vocal proponent of the Edmonton Panel's work, as well as to push for the development of the Energy Transition Strategy. This political support allowed the Edmonton Panel to connect with a focused, existing policy process and provide sufficient public support to punch through the broader petro-state factors otherwise impeding the development of climate policy in Alberta.

Finally, given the strong influence of the energy industry and petro-politics in Alberta, along with associated climate skepticism, the Edmonton Panel worked hard to gather representative interests from across the spectrum of political views and perspectives on climate change. This diversity of interests on the panel was instrumental in bringing conclusions and recommendations from the panel forward to the City of Edmonton. These contextual factors involving demographics and perspectives on climate change in Alberta were recognized by the ABCD project and were a key factor in the design of the deliberations.

The Alberta Energy Efficiency Choices (AEEC) deliberation emerged out of a partnership between ABCD and the Alberta Energy Efficiency Alliance (AEEA). The AEEA is a non-profit stakeholder group interested in advancing energy efficiency within the province (Alberta Energy Efficiency Alliance 2015). AEEC was a province-wide deliberation on energy efficiency, the results of which the AEEA hoped would be helpful in its lobbying efforts with the provincial government (Haas Lyons 2014). In order to appeal to a seemingly hostile provincial government, AEEA felt that it was necessary that the AEEC discussions be as broadly representative as possible. If only a subset of the population, such as Edmontonians or even urban residents, were represented, it would be easier for the government to dismiss whatever findings came out of the discussions. The timing of the AEEC discussions also coincided with what AEEA believed was a policy window. At the time, Alberta was the only jurisdiction in North America without an energy efficiency program, and, according to AEEA executive director Jesse Row, the governing Progressive Conservatives were showing more signs of openness to energy efficiency than to other climate change-related initiatives.

The AEEC deliberation focused on how the government should implement or fund energy efficiency programs. After a brief introduction to concepts such as market transformation, and the various proposals, participants separated into smaller online discussion groups to discuss whether energy efficiency should come through regulations, or through incentives funded by general revenue, additional taxes, or the climate fund paid into through the SGER policy (see chapter 7). AEEA felt this would give them some information about public acceptance regarding certain options with which to approach the provincial government. In keeping with the move toward focused, stakeholder-based public deliberations, this deliberation on energy efficiency is consistent with government interests in supporting tightly organized, focused, and "niche"-oriented public engagement on specific themes. This approach to engagement is low-profile and low-risk for government agencies.

For the Water in a Changing Climate (WCC) project, ABCD partnered with the Oldman Watershed Council (OWC) to conduct a one-day deliberation on climate change issues within a watershed (see chapters 5 and 7). The partnership with the Oldman Watershed Council was not connected to any particular policy moment. This lack of clear policy influence lowered the stakes for the deliberation as a whole. There was no overarching decision or outcome that would encourage opposition from actors hesitant about climate policy or suspicious of public engagement processes. As a result, it served more as a pilot project for both ABCD practitioners and the OWC to explore the process of climate change issue framing at a regional scale and to identify tangible responses and recommendations that are developed by citizens. For ABCD practitioners, this allowed them to experiment with a more organic and open-ended process to frame the issues and set the agenda (Blue 2015). Unlike the Edmonton Panel, which had a much more focused and rigid agenda, the WCC allowed participants to set the range of issues to be discussed. Deliberations also allowed ABCD practitioners insight into the challenge of thinking about meaningful responses to climate challenges in a local and regional context. In this regard, local context and a recent major flood event in southern Alberta conditioned the recommendations from participants. For instance, participants called for "education and information about how to deal with extreme weather events." These outcomes offered insight into the challenges of situating a global challenge like climate change within a regional watershed management context.

Conclusion

This review of public consultation in Alberta speaks to a number of important points in the history of the ABCD project. While the initial vision for ABCD was to convene a high-profile, provincial-level deliberation on climate policy in Alberta, despite several attempts to initiate project activities with provincial counterparts, the project never achieved this vision. Instead, ABCD found fruitful ground for public deliberation at the municipal level in Edmonton with both the Food and Ag Panel and the Edmonton Panel. While ABCD and AEEA did conduct province-wide online discussions about energy efficiency choices, the quality of the deliberations was significantly lower than the municipal deliberations. What the AEEA project achieved with regard to breadth, it sacrificed with regard to depth. Similarly, the WCC deliberation was limited in its quality due to time (one-day) and resource constraints. At the same time, our review of public engagement in Alberta reflects a growing distaste for centralized, high-profile engagements on behalf of the provincial government, and our experience in ABCD is entirely consistent with this shift in public taste. The implications of this trend toward regional, smaller-scale, issue-specific deliberation is taken up in other chapters of this volume in relation to linking public deliberation and system change (see chapters 2 and 8).

Throughout this process, there are a number of key lessons learned. The first is that despite significant institutional and structural forces to the contrary, it is possible to conduct robust deliberative events on topics that are deeply divisive, provided one can find a willing and well-resourced partner with purposeful and well-intentioned links to policy development. The Food and Ag Panel and the Edmonton Panel both fed into ongoing strategic planning processes which provided them with a more tangible connection to the policy-making process. Unfortunately, the Food and Ag Panel did not have the same level of political support enjoyed by the Edmonton Panel, which contributed to the relative success of the latter. The AEEC and WCC deliberations were initiated primarily by ABCD members and partner organizations but had less connection with the policy-making process, fewer resources with which to operate, and no tangible commitment from decision makers on what to do with the recommendations.

A second lesson is how an organization such as ABCD (and other organizations looking at these sorts of processes) defines the success of deliberative events. Some definitions focus on the quality of the deliberations and the influence on participants whereas others define success by substantive changes to policy. With regard to the latter, even the most planned and resourced deliberation ABCD conducted (the Edmonton Panel) had arguably modest influence on the overall development of Edmonton's Energy Transition Strategy. Rather, it provided political support for the policies outlined in the discussion paper (Pembina Institute and HB Lanarc 2012) that fed into the overall Energy Transition Strategy. And yet this contribution should not be diminished. As green deliberative theorist Walter Baber reminds us, "decision processes that are insufficiently democratic are *politically* unsustainable and, therefore, will eventually prove to be *ecologically* unsustainable simply because they will not be able to endure" (Baber 2011, 198). While it may not have substantially altered climate policy in the City of Edmonton, the Edmonton Panel arguably provides the Energy Transition Strategy with significant democratic buttresses to resist stakeholders who might otherwise attempt to weaken the strategy's long-term goals.

Finally, in places like Alberta where public engagement is often little more than public relations, the learning curve for implementing high-quality public deliberation is steep. Demonstrating public deliberation and learning from these experiences with our partners offers an important step toward breaking the old moulds of public consultation in Alberta and offering fresh alternatives.

Notes

1. The authors of this chapter were involved in various ways with ABCD. Geoff Salomons, a PhD candidate in the Department of Political Science at the University of Alberta, served as a research assistant for ABCD from September 2012 until the project's end in 2015. While providing research support for all the ABCD projects, he also served as a table host and note taker for both the Edmonton Citizens' Panel on Climate and Energy Challenges (Edmonton Panel) and the Alberta Energy Efficiency Choices (AEEC) online discussions. John Parkins, a professor in the Department of Resource Economics and Environmental Sociology, was involved in the ABCD project from the outset as a member of the steering committee, as an observer of the Edmonton Panel, and as a supervisor for a graduate student who conducted research with participants of the Edmonton Panel. His work on public deliberation in Alberta predates the ABCD project with attention to public engagement on issues related to natural resource management.

2. Typically, the term "in situ technology" refers to Steam Assisted Gravity Drainage (SAGD), whereby steam is injected underground to melt the viscous bitumen to a more fluid oil and water emulsion, which can then be extracted more easily. The

additional energy required to generate the necessary steam increases the carbon footprint of in situ oil sands when compared with conventional oil sources.

References

Adkin, Laurie. 2014. "Making Climate Change Policy in Alberta." Paper presented at the Canadian Political Science Association, Congress of Humanities and Social Sciences, Brock University, May 29.

———, ed. 2016. *First-World Petro Politics: The Political Ecology and Governance of Alberta.* Toronto: University of Toronto Press.

Alberta Energy. 2017. *Resource Revenue Workbook.* Accessed January 25. http://www. energy.alberta.ca/Org/docs/Revenueworkbook.xls.

Alberta Energy Efficiency Alliance. 2015. Accessed January 25, 2017. http://www.aeea.ca/.

Andrais, Jim. 2015. Telephone Interview with Lorelei Hanson, December 9.

Angus Reid Institute. 2015. "Most Canadians Support Carbon Pricing; but Less Consensus on Effectiveness of Such Measures." Accessed January 25, 2017. http:// angusreid.org/carbon-pricing/.

Baber, Walter F. 2011. "Review: Environmental Politics and Deliberative Democracy: Examining the Promise of New Modes of Governance." *Journal of Environmental Policy and Planning* 13(2): 197–99.

Blue, Gwendolyn. 2015. "Framing Climate Change for Public Deliberation: What Role for Interpretive Social Sciences and Humanities?" *Journal of Environmental Policy and Planning* 18(1): 67–84.

British Petroleum. 2017. *BP Statistical Review of World Energy.* Accessed January 25. http://www.bp.com/en/global/corporate/energy-economics/statistical-review-of-world-energy.html.

Centre for Public Involvement. 2012. *Session Outputs and Schedule.* Unpublished Meeting Schedule. April 15. Faculty of Extension, University of Alberta.

City of Edmonton. 2011. *The Way We Green: The City of Edmonton's Environmental Strategic Plan.* Accessed January 27, 2015. http://www.edmonton.ca/city_government/documents/PDF/TheWayWeGreen-approved.pdf.

———. 2012. *City-Wide Food and Urban Agriculture Strategy Report on Citizen Panel Process and Recommendations.* Accessed January 25, 2017. http://www.edmonton.ca// documents/PDF/Food_and_Ag_Strategy_Citizen_Panel_Report_Sept_2012.pdf.

Curry, Bill, and Shawn McCarthy. 2011. "Canada Formally Abandons Kyoto Protocol on Climate Change." *Globe and Mail.* Accessed January 25, 2017. http://www. theglobeandmail.com/news/politics/canada-formally-abandons-kyoto-protocol-on-climate-change/article4180809/.

Davidson, Deborah J., and Norah A. MacKendrick. 2004. "All Dressed Up with Nowhere to Go: The Discourse of Ecological Modernization in Alberta, Canada." *Canadian Review of Sociology* 41(1): 47–65.

Dryzek, John S. 1995. "Political and Ecological Communication." *Environmental Politics* 4(4): 13–30.

Dyer, Simon, Matthew Bramley, Marc Hout, and Matt Horne. 2011. "Responsible Action? An Assessment of Alberta's Greenhouse Gas Policies." Pembina Institute. Accessed January 27, 2015. http://www.pembina.org/pub/2295.

EnviroEconomics. 2009. "Act Locally The Municipal Role in Fighting Climate Change." *Federation of Canadian Municipalities.* Accessed January 25, 2017. https://www.fcm.ca/Documents/reports/Act_Locally_The_Municipal_Role_in_Fighting_Climate_Change_EN.pdf.

Environics Institute for Survey Research and the David Suzuki Foundation. 2014. "Focus Canada 2014: Canadian Public Opinion About Climate Change." Accessed January 25, 2017. http://www.davidsuzuki.org/publications/downloads/2014/Focus%20Canada%202014%20-%20Public%20opinion%20on%20climate%20change.pdf.

Environment Canada. 2014. "Canada's Emissions Trends." *Environment Canada.* Accessed January 25, 2017. https://ec.gc.ca/ges-ghg/E0533893-A985-4640-B3A2-008D8083D17D/ETR_E%202014.pdf.

Forum Research Inc. 2014. "Vast Majority Accept Climate Change." *The Forum Poll.* Accessed January 27, 2015. http://poll.forumresearch.com/post/99/vast-majority-accept-climate-change/.

FCM (Federation of Canadian Municipalities). 2015. "Green Municipal Fund." Accessed January 27. http://www.fcm.ca/home/programs/green-municipal-fund.htm.

Gifford, Robert. 2011. "The Dragons of Inaction: Pychological Barriers That Limit Climate Change Mitigation and Adaptation." *American Psychologist* 66(4): 290–302.

Government of Alberta. 2003. "Water for Life: Alberta's Strategy for Sustainability." Accessed January 27, 2015. http://aep.alberta.ca/water/programs-and-services/water-for-life/strategy/documents/WaterForLife-Renewal-Nov2008.pdf.

———. 2008. "Water for Life: A Renewal." Accessed January 27, 2015. http://aep.alberta.ca/water/programs-and-services/water-for-life/strategy/documents/WaterForLife-Renewal-Nov2008.pdf.

———. 2009. "Terms of Reference for Developing the Lower Athabasca Regional Plan." Accessed January 27, 2015. http://www.ceaa.gc.ca/050/documents/45582/45582E.pdf.

———. 2015a. "Competitive Corporate Taxes." Accessed January 27, 2015. http://www.albertacanada.com/business/overview/competitive-corporate-taxes.aspx.

———. 2015b. "Highlights of the Alberta Economy." Accessed January 27, 2015. http://www.albertacanada.com/files/albertacanada/SP-EH_highlightsABEconomyPresentation.pdf.

———. 2016. "Tax and Revenue Administration – Links for Information Provincial Sales Tax and Harmonized Sales Tax." Accessed January 27. http://www.finance. alberta.ca/publications/tax_rebates/provincial_sales_tax_links.html.

Government of Canada. 2015. "Archived – Government of Canada Announces 2030 Emissions Target." Accessed January 27, 2015. http://news.gc.ca/web/article-en. do?nid=974959.

Haas Lyons, Susanna. 2014. Learning from Technology's Role in Energy Efficiency Choices. Unpublished Paper, Alberta Climate Dialogue 2014 Workshop.

Hanson. Lorelei L., and Deborah Schrader. 2016. "Identifying Opportunities and Hurdles for Food Security: A Critical Examination of the City of Edmonton's Food and Agriculture Strategy." *Interdisciplinary Environmental Review* 17(2): 98–115.

Harrison, Kathryn. 2012. "A Tale of Two Taxes: The Fate of Environmental Tax Reform in Canada." *Review of Policy Research* 29(3): 383–407.

Harrison, Kathryn, and Lisa McIntosh Sundstrom, eds. 2010. *Global Commons, Domestic Decisions: The Comparative Politics of Climate Change*. Cambridge: MIT Press.

Homer-Dixon, Thomas. 2013. "The Tar Sands Disaster." *New York Times*, April 1. Accessed January 25, 2017. http://www.nytimes.com/2013/04/01/opinion/the-tar-sands-disaster.html?_r=1.

IPPC (Intergovernmental Panel on Climate Change). 2014. *Fifth Assessment Report*. Accessed January 25, 2017. http://www.ipcc.ch/report/ar5/.

ICLEI (International Council for Local Environmental Initiatives). 2017. Accessed January 25, 2017. http://www.icleicanada.org/.

Karl, Terry Lynn. 1997. *The Paradox of Plenty: Oil Booms and Petro-States*. Berkeley: University of California Press.

Kasperson Roger E. 2006. "Rerouting the Stakeholder Express." *Global Environmental Change* 16(4): 320–22.

Kaufman, Jerome L. 2009. "Food System Planning: Moving Up the Planner's Ladder." *Plan Canada* 49(2): 12–16.

KlineGroup. 2008. From Oil Sands to a World-Class Eco-Industrial Chemical Cluster for the Greater Edmonton Area. Accessed January 25, 2017. http://www.energy. alberta.ca/Org/pdfs/HUFTosClusterStudyMay08.pdf.

Leach, Andrew. 2013. "Canada, the Failed Petrostate?" *Maclean's Magazine*, November 4. Accessed January 25, 2017. http://www.macleans.ca/economy/economicanalysis/canada-the-failed-petrostate/.

Lindblom, Charles E. 1977. *Politics and Markets: The World's Political Economic Systems*. New York: Basic Books.

Mitchell, Timothy. 2011. *Carbon Democracy: Political Power in the Age of Oil*, 2nd ed. Brooklyn: Verso.

Oliver, Joe. 2012. An Open Letter from the Honourable Joe Oliver, Minister of Natural Resources. *Natural Resources Canada*. Accessed January 27, 2015. https://www.nrcan. gc.ca/media-room/news-release/2012/1/1909.

Paris, Max. 2013. "Energy Industry Letter Suggested Environmental Law Changes." *CBC News*, January 9. Accessed January 27, 2015. http://www.cbc.ca/news/politics/ energy-industry-letter-suggested-environmental-law-changes-1.1346258.

Parkins, John R. 2006. "De-Centering Environmental Governance: A Short History and Analysis of Democratic Processes in the Forest Sector of Alberta, Canada." *Policy Sciences* 39(2): 183–203.

———. 2011. "Deliberative Democracy, Institution Building, and the Pragmatics of Cumulative Effects Assessment." *Ecology and Society* 16(3): 20.

Parkins, John R., and A. John Sinclair. 2014. "Patterns of Elitism Within Participatory Environmental Governance." *Environment and Planning C: Government and Policy* 32(4): 746–761.

Pembina Institute and HB Lanarc. 2012. Edmonton's Energy Transition: Discussion Paper. Accessed January 23, 2017. https://www.edmonton.ca/city_government/ documents/PDF/Edmonton_Energy_Transition_Discussion_Paper.pdf.

Prime Minister of Canada's Office. 2016. Communiqué of Canada's First Ministers. Accessed May 12. http://pm.gc.ca/eng/news/2016/03/03/communique-canadas-first- ministers.

Rachlinski, Jeffrey J. 2000. "The Psychology of Global Climate Change." *University of Illinois Law Review* 1: 299–320.

Robinson, Pamela J., and Christopher D. Gore. 2005. "Barriers To Canadian Municipal Response To Climate Change." *Canadian Journal of Urban Research* 14(1): 102–20.

Ross, Michael L. 2012. *The Oil Curse: How Petroleum Wealth Shapes the Development of Nations*. Princeton: Princeton University Press.

Ryan, David L. 2013. *Boom and Bust Again: Policy Challenges for a Commodity-Based Economy*. Edmonton: University of Alberta Press.

Sadler, Barry. 1977. "Basic Issues in Public Participation: A Background Perspective." In *Involvement and Environment: Proceedings of the Canadian Conference on Public Participation*, edited by Barry Sadler, 1–12. Edmonton: Environment Council of Alberta.

Salomons, Geoff, and George Hoberg. 2014. "Setting Boundaries of Participation in Environmental Impact Assessment." *Environmental Impact Assessment Review* 45: 69–75.

Satterthwaite, David. 2009. "The Implications Of Population Growth And Urbanization For Climate Change." *Environment and Urbanization* 21(2): 545–67.

Schneider, R.R. 2001. "Whatever Happened to the Alberta Forest Conservation Strategy?" *Encompass* 5: 9–13.

Seawright, Jason, and John Gerring. 2008. "Case Selection Techniques in Case Study Research: A Menu of Qualitative and Quantitative Options." *Political Research Quarterly* 61(2): 294–308.

Shrivastava, Meenal, and Lorna Stefanick. 2012. "Do Oil and Democracy Only Clash in the Global South? Petro Politics in Alberta, Canada." *New Global Studies* 6(1): 1–27.

———, eds. 2015. *Alberta Oil and the Decline of Democracy in Canada*. Edmonton: Athabasca University Press.

Solomon, Evan, and Kristen Everson. 2014. "7 Environmental Charities Face Canada Revenue Agency Audits." Accessed January 27, 2017 http://www.cbc.ca/news/politics/7-environmental-charities-face-canada-revenue-agency-audits-1.2526330.

Statistics Canada. 2016. "Table 380-0064 Gross Domestic Product, Expenditure-Based." Accessed January 27, 2017 http://www5.statcan.gc.ca/cansim/a26?lang=eng&retrLang=eng&id=3800064&&pattern=&stByVal=1&p1=1&p2=-1&tabMode=dataTable&csid=.

Stefanick, Lorna, and Kathleen Wells. 2000. "Alberta's Special Places 2000: Conservation, Conflict, and the Castle-Crown Wilderness." In *Biodiversity in Canada: Ecology, Ideas and Action*, edited by Stephen Bocking, 367–90. Peterborough, ON: Broadview Press.

Synergy Alberta. 2016. "Vision and Mission." Accessed January 25, 2017. http://www.synergyalberta.ca/vision-and-mission.

Taft, Kevin, Melville L. McMillan, and Junaid Jahangir. 2012. *Follow the Money: Where Is Alberta's Wealth Going?* Calgary: Detselig.

Waller, Lori Theresa. 2012. Oil Servitude and the New Canadian Petrostate: An Interview with Andrew Nikiforuk. *Rabble*, October 29. Accessed January 25, 2017. http://rabble.ca/news/2012/10/oil-servitude-and-new-canadian-petrostate-interview-andrew-nikiforuk.

Weber, Elke U. 2011. "Psychology: Climate Change Hits Home." *Nature Climate Change* 1(1): 25–26.

Wesley, Jared J. 2011. *Code Politics: Campaigns and Cultures on the Canadian Prairies*. Vanouver: University of British Columbia Press.

Beyond the Usual Suspects

Representation in Deliberative Exercises

Shelley Boulianne

Public deliberation exercises are intended to provide more inclusive forums for policy debates, in contrast to elite-dominated approaches to public consultation. Their legitimacy is, in part, derived from a participant selection process that is representative of the broader public (Fournier et al. 2011, 148). However, if these exercises are intended to replace elite-dominated approaches, then they should also be judged in terms of the degree to which they achieve demographic and attitudinal diversity. Ryfe and Stalburg (2012, 54) argue that "the question of who deliberates represents one of the most significant gaps in our understanding of deliberative practices." Without examining those involved in a public deliberation exercise, it is difficult to evaluate whether it reaches its goals of inclusiveness and representativeness.

Deliberation organizers use a variety of strategies to establish the representativeness of the participants in their public deliberation exercises. This chapter considers representation in deliberative exercises as the degree to which there is a match between the participants in a deliberative exercise and the broader public as established by a census or other high-quality survey. The minimum standard for demographic representation is based on the census profile for the geographic area which compares age, gender, and education of participants in relation to the population.

More recently, scholars have opted to go beyond demographic representation and compare the group's attitudinal composition to their citizen counterparts as determined by public opinion polls. The focus on attitudes raises issues of inclusiveness, as certain segments of the population may have different perspectives on the topic being deliberated. Engaging these disparate viewpoints is critical for making the consultation deliberative. Furthermore, the inclusion of minority opinions is important because these voices are often systematically excluded from the policy-making process.

This chapter highlights four different deliberative exercises around the topic of climate change that involved members of Alberta Climate Dialogue (ABCD). For each deliberative exercise, I consulted on the design of participant surveys. I advised on the wording of demographic questions to ensure comparability with Statistics Canada approaches and to ensure consistency across the deliberative exercises. For the Citizens' Panel on Edmonton's Energy and Climate Challenges, I helped design recruitment materials and the Interactive Voice Response (IVR) survey used to assess broader public opinion about climate change. I will discuss the approaches I used as well as other innovative approaches used to recruit citizens to participate in public deliberations. The four recruitment approaches illustrate the challenges of ensuring representativeness and inclusiveness in deliberations about climate change. In deciding between recruitment approaches, deliberation organizers need to recognize the trade-offs between representativeness and inclusiveness.

Climate change is a particularly difficult policy issue given the uncertainty around the impacts as well as the need for both global and localized responses. In this context, citizen engagement is not only a challenge but a necessity.

Recruitment Approaches

Most typologies of recruitment approaches focus on the distinction between random sampling and self-selected samples (e.g., Mao and Adria 2013; Ryfe and Stalburg 2012). However, these categorizations falsely dichotomize the two recruitment approaches, in that they ignore the self-selection process that occurs within the process of random sampling. For example, in all three Citizens' Assemblies on Electoral Reform (convened in Ontario, British Columbia, and the Netherlands) citizens were randomly chosen from voter registration lists. Of the participants chosen from the voter registration list, only 6 to 7 per cent expressed an interest in participating in the deliberative event (Fournier et al. 2011, 32). Although these projects used random sampling, they acknowledged the role of self-selection in recruiting participants. Each step in the recruitment process, including volunteering to participate, showing up to participate, and attending all meetings involves some self-selection (see Griffin et al. 2015), which compromises the idea of random sampling. Instead of focusing on random sampling issues versus self-selection, this chapter categorizes recruitment approaches in terms of the goals of achieving demographic representation (representativeness) and attitudinal diversity (inclusiveness).

Demographic Representation

Most deliberative events opt for demographic representation based on the census profile for the geographic area (Gastil 2000). This strategy may or may not involve random sampling. For example, America*Speaks*, a Washington, DC–based non-profit focused on citizen engagement in public decision making, used self-selected samples for their more than forty-five deliberative 21st Century Town Meetings (Lukensmeyer and Brigham 2005). These deliberative exercises addressed policy issues ranging from Social Security to regional planning, such as rebuilding the World Trade Center site. The organization also undertook targeted recruitment in areas of expected under-representation, for example recruiting seniors and youth in deliberations about Social Security (Lukensmeyer and Brigham 2005). In most cases, the goal was to reflect the demographic composition of the region (Lukensmeyer and Brigham 2005).

Demographic diversity can also be achieved by employing random selection. For example, Farrar et al. (2010) recruited New Haven and area residents for a deliberation about airport expansion. They compared their deliberative participants to the voting population in terms of age, gender, marital status, education, income, and race (Farrar et al. 2010). Similarly, for the World Wide Views global citizen consultation project, the Danish Board of Technology encouraged countries to organize their deliberative exercises to ensure representation based on age, gender, occupation, education, and geography (Blue 2012). Focusing on demographic representation is the most popular technique for establishing the representation of deliberative groups (Hobson and Niemeyer 2011).

Despite efforts to establish representation based on demographic variables, deliberative exercises consistently fail to attract particular groups of people. Education, gender, and age are most commonly discussed in demographic representation (Andersen and Hansen 2007; Farrar et al. 2009; Farrar et al. 2010; Fishkin et al. 2010; French and Laver 2009; Griffin et al. 2015; Hall, Wilson, and Newman 2011; Hansen and Andersen 2004; Hobson and Niemeyer 2011; Setälä, Grönlund, and Herne 2010; Strandberg and Grönlund 2012). However, deliberative events tend to over-represent men, under-represent young people, and almost consistently over-represent the educated. While some deliberative exercises have addressed gender equity and have had some success with age group representation, representation based on education remains the greatest challenge to demographic representation in deliberative exercises (Farrar et al. 2010; Fournier et al. 2011; French and Laver 2009; Merkle 1996). In the BC and Ontario Citizens' Assemblies on Electoral Reform, within the deliberative group 44 per cent held university degrees, whereas 19 to 20 per cent of the population as a whole have university degrees (Fournier et al. 2011). Farrar et al. (2010) found that approximately one-third of their participants had graduate degrees, whereas in the broader voting public, only 12 per cent have graduate degrees. These findings point to a consistent pattern of over-representation of educated people in deliberative events.

Fournier et al. (2011) provide two counter-arguments to concerns about demographic bias. First, they argue that their deliberative participants are far more representative than legislative assemblies. The participants are "expected to have preferences that are more congruent with those of the general population than those of elected politicians" (Fournier et al. 2011, 54). Second, they examine whether there are differences in policy preferences based on education. The implication is that educational differences in policy preferences would compromise the legitimacy of the deliberative body. However, they used a public opinion poll and found no education-based differences in opinions about the policies being examined and conclude that "the effect of a more representative assembly would thus have been small" (Fournier et al. 2011, 61). However, if the deliberative body is being compared to a public opinion poll, how do we know that the poll respondents were representative of the broader public? Hall, Wilson, and Newman (2011) compared the demographic composition of their deliberative participants, poll results, and census data and found that the deliberative body's composition was similar to that of the poll respondents, but both were dissimilar to the census profile for the region. These findings suggest that deliberative exercises replicate the bias of polls rather than offering a more inclusive form of public participation.

Demographic representation is, in some respects, at odds with principles of inclusion. Recruiting social groups in proportion to their representation in the population would replicate minority statuses which exist in the population. Instead, there could be value in oversampling particular groups whose views may not be represented in the typical policy-making process (Blue 2012).

Oversampling this group would help ensure "a critical mass of participants from minority social groups . . . to ensure their voices are recognized and heard" (Bächtiger, Setälä, and Grönlund 2014, 230). For example, French and Laver (2009) oversampled citizens who live in an electoral division hosting a proposed waste treatment facility, which was the subject of the deliberation. This sampling approach ensures that their deliberative participants include those with "local knowledge" (French and Laver 2009, 428). James (2008, 108) also argues in favour of oversampling groups who are disproportionately affected by the policy domain; this recruitment approach may increase access to "distinct forms of social knowledge more likely to be found among members of such groups." As another example, the Canadian edition of the World Wide Views project oversampled Indigenous and northern people for a deliberation on climate change (Blue 2012). The assumption is that Indigenous and northern people are differentially affected by climate change and have alternative knowledge about the issue (Blue 2012). Indeed, some argue that the value of deliberative exercises, as opposed to other forms of engagement, is the inclusion of groups who would not have a voice otherwise (Blue 2012; Karjalainen and Rapeli 2015). As such, replicating minority status, which adheres to traditional principles of representation, in deliberative exercises would be counterproductive to the goal of inclusion.

Attitudinal Diversity

Demographic diversity is often used as a proxy for attitudinal diversity, and in many cases, this logic is clearly flawed, particularly when the demographic variables focused upon do not predict attitudinal differences related to the topic of deliberation. Instead, the issue of representation would be better addressed by ensuring attitudinal diversity, particularly on the topic of deliberation. Gastil, Knobloch, and Kelly (2012, 224–25) write:

> With regard to representativeness, the final body of citizens who attend the event . . . should be surveyed to determine their relevant demographic and ideological (attitudinal) characteristics. These characteristics can then be compared against relevant census and survey data for the targeted geographic/political region.

James (2008) encourages organizers to consider which demographic variables predict differences in opinions on the policy matter and which groups will benefit more than others from a particular policy direction. In the case of the BC Citizens' Assemblies, visible minority participants had different preferences about electoral reform than other citizens (James 2008). This group's under-representation could undermine the legitimacy of the deliberating body (James 2008). When it comes to climate change, similar questions could be asked: Which demographic variables affect differences in policy preferences and who benefits more from the different policy proposals?

Bächtiger, Setälä, and Grönlund (2014, 231) recognize the value of attitudinal diversity but also the challenge of recruiting on attitudes about "scientifically complex issues on which people might not have clear pre-deliberation opinions." This concern is particularly relevant for recruitment for climate change deliberations. If the average citizen's knowledge level is low, then recruitment strategies need to be more cognizant of the potential for bias. Citizens who are more knowledgeable about the topic may self-select to participate, leaving those with minimal knowledge excluded from the deliberation. What distinguishes a public deliberation from a stakeholder consultation is the inclusion of non-experts (Blue and Medlock 2014). Recruiting for climate change deliberations is particularly difficult because the framing is often tied to science and can thus restrict knowledge claims to those made by scientific experts (Blue and Medlock 2014). This framing can alienate those without advanced education in the sciences. Furthermore, the experiences of climate change can be elusive as a personal or perceptible experience (Weber 2010; Weber and Stern 2011). Blue and Medlock (2014, 6) write that "Greenhouse Gas (GHG) emissions, for instance, are imperceptible to the senses without the assistance of science and technology." In the context of climate change policy, participants may not have preconceived notions about climate change or how to address it. As such, ensuring attitudinal diversity could be difficult.

Scholarship tends to focus on attitudinal variables that predict political engagement. For example, many scholars compare their deliberative group to public opinion data regarding political interest, efficacy, confidence, and political knowledge (Fournier et al. 2011; French and Laver 2009; Griffin et al. 2015; Grönlund, Setälä, and Herne 2010; Hansen and Andersen 2004; Merkle 1996; Luskin, Fishkin, and Jowell 2002; Strandberg and Grönlund 2012). Every one of these studies documents that the deliberative participants do not represent the public on at least one of these attitudinal variables. Deliberative participants tend to be more politically interested, efficacious, and knowledgeable than the broader public, as established by public opinion polls. To address concerns about the bias, Luskin, Fishkin, and Jowell (2002, 466) argue that few of the differences are statistically significant and the differences are "fairly modest." They also argue that "ordinary polls generally possess the same sort of bias" (Luskin, Fishkin, and Jowell 2002, 466). Their argument accentuates, rather than allays, concerns about representation. Again, in this context, deliberative exercises replicate the bias of polls rather than offering a more inclusive form of public participation.

A smaller set of studies has examined how deliberative participants compare to non-participants on attitudes related to the deliberative topic. Hall, Wilson, and Newman (2011) compared a public opinion poll of 504 respondents to their sixty-two event participants and documented differences in levels of environmental concern and beliefs in the environmental harm of fossil fuels. Participants in the deliberative exercise about energy issues in Idaho "had high interest and pre-existing attitudes about energy issues" (Hall, Wilson, and Newman 2011, 9). Andersen and Hansen (2007) compared poll respondents to participants in a deliberation about adopting the euro. Those people recruited to participate in the deliberation had planned to vote yes and were less likely to be "undecided" than poll respondents (Andersen and Hansen 2007, 536, Table 2). Comparing a lengthy list of funding projects, Fishkin et al. (2010) examined policy opinions for the 235 Chinese citizens who participated in the deliberation and those who completed the poll but did not participate (n=34). They found only one statistically significant difference (a 21 percentage point difference) between the two groups, but several other differences were quite large (Fishkin et al. 2010). They found significant differences between the demographic (age, gender, education, and occupation) composition of the participants and those who did not participate (Fishkin et al. 2010).

The most comprehensive and serious treatment of opinion bias is a study by Karjalainen and Rapeli (2015). Opinions about the deliberation topic played a key role in whether participants showed up to the deliberation (Karjalainen and Rapeli 2015). They found that in a deliberation about immigration, those who opposed immigration were under-represented on deliberation day (Karjalainen and Rapeli 2015). All of these findings were post–data collection reflections. Attitudinal diversity was not a guiding principle for the recruitment strategy. A more innovative recruitment approach would be to assess attitudinal diversity during the recruitment stages and adjust recruitment strategies to ensure a proper reflection of attitudes prior to the deliberative event.

Recruitment Techniques

The most common recruitment techniques are to post advertisements in local newspapers, libraries, and other public spaces, to send invitations to thousands of citizens with the hopes that sufficient numbers will respond to the invitation, or to concurrently conduct a public opinion poll and recruitment for the deliberative exercise. These different strategies have different claims to representativeness and inclusiveness. While polls may be representative, they may not be inclusive, since citizens are randomly recruited with little consideration as to the unique perspectives that exist within subpopulations.

The cheapest form of recruitment relies on a self-selection process. Advertisements are posted at libraries and other public spaces asking for volunteers to participate. Alternatively, the advertisement can be placed on websites, in local newspapers, or distributed via electronic mailing lists. In targeted recruitment campaigns, these strategies are also used, but the recruitment campaigns focus on specialized newspapers or websites that target specific population groups. Another targeted recruitment approach involves contacting organizations that represent or serve specialized populations. The organization often forwards or posts recruitment messages on behalf of the deliberation organizers. All of these techniques rely on a self-selection process. These recruitment strategies emphasize inclusiveness by recognizing that some groups are differentially impacted by policy approaches or may have unique viewpoints about the topic being deliberated upon. The self-selection process likely produces a group of citizens who are highly interested in the topic of discussion.

The other recruitment techniques are clearly aligned with public opinion polling. One approach is to send out recruitment packages to thousands of citizens with the hopes that a sufficient number will return their forms expressing interest in participation. The response rate to these invitations would make most public opinion researchers cringe. For example, Strandberg and Grönlund (2012) sent out invitations to 6,000 Finnish people, 147 volunteered, and only 79 actually participated. In the Canadian arm of the World Wide Views project, 3,000 invitations were sent and 98 people responded to express their interest in participating (Blue 2012).

Another popular approach is to engage in a public opinion poll, which concludes with a question about interest in participating in a deliberative exercise (e.g., Fishkin et al. 2010; French and Laver 2009; Hansen and Andersen 2004). With these designs, researchers can compare the demographic and attitudinal composition of poll respondents to those who agree to participate in the deliberative exercise. Not only does this approach establish representativeness, but the poll respondents can serve as a control group for comparison (French and Laver 2009). While this research design is one of the stronger methodological approaches to recruitment, descriptions of these methods tend to be uncritical of the self-selection inherent in this process. Karjalainen and Rapeli (2015) highlight the layers of self-selection, and possible bias, introduced into a process that involved contacting almost 12,000 Finnish people, but having only 200 people show up to deliberate. Employment status is a key driver in whether or not people are willing to participate (Karjalainen and Rapeli 2015; Neblo et al. 2010). If employment status affects viewpoints about the topic of deliberation, then this bias could detrimentally impact the representativeness of the deliberating group of citizens. In general, recruitment techniques that replicate public opinion polling techniques have stronger claims to representativeness.

Regardless of the recruitment technique, deliberative exercises often involve some kind of incentive or honorarium for participation. The use of incentives follows best practices in focus group recruitment. Experts suggest that face-to-face focus groups of two hours should be accompanied by incentives of at least $50 (Stewart, Shamdasani, and Rook 2011). However, adjustments should be made to accommodate the costs of travel and childcare needs related to participation (Stewart, Shamdasani, and Rook 2011).

The following sections describe four deliberative exercises in which ABCD participated that used a variation of the above techniques to recruit participants to deliberate on a topic related to climate change. These case studies highlight the challenges of achieving representativeness while ensuring inclusiveness. For each case study, the demographic profile of the recruited participants is discussed in relation to population characteristics. In some cases, attitudinal comparisons are also made to discuss the success of the recruitment strategy. In each case, the advantages and disadvantages, including costs, are listed.

Case Study 1: City-Wide Food and Urban Agriculture Citizen Panel

This deliberative group was organized by the Centre for Public Involvement in partnership with the City of Edmonton's Sustainable Development department, and involved five members of ABCD in minor roles of assisting with various research activities (see chapter 1). The citizens met six times over two months in the spring of 2012 to discuss, and provide input into *fresh*—Edmonton's food and urban agriculture strategy. The meetings included two full-day and four half-day sessions (see chapter 1). Participants were offered $150 for participating in the deliberation. Of the sixty-six participants recruited, forty-four were enlisted through random digit dialing and the remaining participants found through community groups, universities, and lists of known volunteers (personal communication, Fiona Cavanagh, August 5, 2015). The targeted recruitment was successful in including an appropriate representation of youth and visible minorities as well as two low income people and three people who did not speak English (translation services were provided) (Fiona Cavanagh, email message to author, August 6, 2015). Fifty-eight panelists participated (City of Edmonton 2012).

The goal in recruiting participants for this panel was to ensure a diversity of participants with respect to gender, length of residence in Edmonton, Indian status, visible minority status, disability status, and city ward of residence (personal communication, Fiona Cavanagh, August 5, 2015). However, women were over-represented in terms of those who participated in the deliberative exercise (City of Edmonton 2012). While population estimates suggest that 5 per cent of Edmonton residents have an Aboriginal identity (Statistics Canada 2011), only 2 per cent of the panel identified as such (City of Edmonton 2012). The recruitment strategy sought representation from visible minorities and was successful in achieving representation comparable to the census profile for the city (City of Edmonton 2012; Statistics Canada 2011). In terms of disability, the recruitment strategy failed to match the census profile (Statistics Canada 2015).

Disadvantage

The recruitment through interviewer-led phone calls was conducted by graduate students. This group was expensive to employ and required a good deal of specialized training. Despite training efforts, there were inconsistencies in recruitment practices from recruiter to recruiter. The labour cost for the recruiters was approximately $6,700 for the forty-four participants who were randomly recruited (personal communication, Fiona Cavanagh, August 5, 2015). This estimate does not include the labour involved in targeted recruitment to the various community agencies and universities.

The hybrid recruitment approach produced an over-representation of educated people. Statistics Canada (2011) estimates that 38 per cent of the Edmonton population have high school or less education, whereas the deliberative group included only 6 per cent of this education group. Despite the concerted efforts to ensure diversity, the participating group was largely composed of university graduates.

While efforts were made to ensure demographic diversity, the recruitment process did not include a general population, probability-based survey, which would have enabled a comparison of the attitudinal diversity of participants compared to the general population. This is a particular concern when recruitment is conducted through community groups. Members of community groups may be like-minded and thus may not ensure a diverse range of attitudes about the topic being deliberated upon. A survey of participants suggests that there were more left-leaning participants than right-leaning participants, although the most common response was "middle of the road" (City of Edmonton 2012). This ideological representation could be explained by the recruitment of university students, since university students tend to be more left-leaning (Olcese, Saunders, and Tzavidis 2014). Studies show that views about climate change are driven by ideological orientation (Davidson and Haan 2012). As such, ideological bias could undermine the work of a deliberative body. While the panel included a range of ideologies, a general population survey was not available to establish how the panel compared to broader public opinion.

Advantage

The strategy of random recruitment with targeted recruitment was successful in ensuring proportional representation for members of visible minorities and young people. Approximately 28 per cent of the Edmonton population consists of people aged 18 to 34 years and 29 per cent of panelists were in this age group (City of Edmonton 2012; Statistics Canada 2012). The inclusion of this age group is important, as young people's future well-being may depend on successful climate change policies. The strong representation of young people reflected targeted recruitment at universities. The successful recruitment of visible minorities was in part attributable to targeted recruitment through community organizations. This recruitment approach has some clear success in achieving inclusiveness.

The random recruitment part of the approach addressed concerns about representativeness. The forty-four participants who were randomly recruited could, arguably, serve as a representative body of Edmontonians. While there were biases noted in the composition of the entire deliberative group, it is unclear whether the bias was introduced by the targeted recruitment or through the random recruitment process. A comparison of the two groups would help advance research in this area.

Case Study 2: Citizens' Panel on Edmonton's Energy and Climate Challenges

The Citizens' Panel on Edmonton's Energy and Climate Challenges was organized in partnership with ABCD, the Centre for Public Involvement, and the City of Edmonton's Office of the Environment. Citizens were recruited to participate in a six-Saturday event to learn about the city's energy and climate challenges and to provide policy recommendations related to these topics. The recruitment strategy for this panel was designed based on successes and challenges in the 2012 Food and Urban Agriculture Citizen Panel.

Addressing concerns about attitudinal diversity, the recruitment process included a probability-based random sample of citizens to assess their views about climate change and other related views. The intention of this practice was to ensure attitudinal representation. In addition, the Centre for Public Involvement redesigned its recruitment materials and practices and contracted a third party to conduct recruitment (personal communication, Fiona Cavanagh, August 5, 2015). Probit, a subsidiary of EKOS Research, conducts ongoing recruitment of citizens, via interactive voice response surveys (IVR), to participate in its online panel. These IVR surveys include landline and cellphone-only households.

When citizens are contacted, they are asked a small number of survey questions, and are then asked if they would like to participate in further research. Of those who agreed to participate in further research, 2,400 people were re-contacted by phone to complete a recruitment survey (CPEECC 2013). Of these participants, more than 300 citizens expressed interest and availability to participate in the Citizens' Panel in fall 2012 (CPEECC 2013). They were told that if they were selected they would receive approximately $400 as an honorarium (with some adjustments for childcare and transportation as well as regular attendance at meetings) (Fiona Cavanagh, email message to author, September 16, 2014). Participants were then asked to complete an informed consent form for research purposes and a Freedom of Information and Privacy Protection consent form, which would allow their names, contact information, and survey responses to be shared with the Centre for Public Involvement and

Alberta Climate Dialogue. After multiple contacts, 101 citizens returned the signed consent form (Elliott Gauthier, email message to author, July 10, 2015).

Upon review of the age profile of these 101 citizens, the Centre for Public Involvement engaged in targeted outreach to try to obtain representation from young people aged eighteen to twenty-nine years (CPEECC 2013). In total, sixty-six citizens were selected to ensure quotas were met to match the population distribution in terms of age, gender, education, ethnicity, disability, households with children, household income, city ward, and households in which a member is employed by the energy sector (CPEECC 2013). The panel mimicked census data for the city on gender, age, and household size, but under-represented households with children in the home (Statistics Canada 2012). In terms of education, those with high school or less were under-represented (29 per cent) in proportion to their representation in the Edmonton population (38 per cent) (Statistics Canada 2011). However, of the four case studies, this project had the greatest success with the recruitment of this education group.

Probit also conducted a separate IVR survey to establish broader public opinion on key attitudinal variables. In terms of attitudinal variables, the panelists were more efficacious and trusting, and liked living in Edmonton more than respondents to the public opinion poll (CPEECC 2013). Panelists were slightly more likely to believe that climate change is happening and that climate change is caused by humans (CPEECC 2013). The panelists were also more likely to pay attention to energy and climate issues and more likely to view governments, industry, and individual citizens as having a greater role to play in addressing climate change (CPEECC 2013). While this public opinion data was useful in assessing attitudinal diversity, the data presents a challenge in trying to determine which attitudes to focus upon to ensure representation. In the end, priority was given to beliefs about the existence of climate change.

Disadvantage

This recruitment approach was expensive. Costs were associated with the recruitment of participants and administration of the IVR survey to establish the broader public's attitudes related to the deliberation topic. The total cost for both initiatives was approximately $13,000 (CPEECC 2013). The attitudinal survey was conducted as an IVR survey, which is substantially cheaper than a telephone survey. While this mode reduced the costs of data collection, the trade-off was a low response rate (Loptson and Boulianne 2013).

Also, some aspects of the recruitment process fell short of achieving their intended goals. The IVR survey targeted both cellphone and landline numbers; the inclusion of cellphone-only lines was expected to address the challenge of recruiting young people, but this recruitment approach failed to produce sufficient representation of young people. As such, in addition to the high costs of recruitment, additional targeted recruitment strategies were required to address these deficiencies. In other words, the random recruitment process was not successful in ensuring proper representation based on age.

Despite the demographic diversity of the panel, the panelists were not representative of the broader public on attitudes related to climate change and who is responsible for addressing climate change (CPEECC 2013). As such, demographic diversity does not ensure attitudinal diversity. The group of deliberative participants was more likely to believe in climate change, report higher efficacy and trust, and enjoy living in Edmonton, than the broader public (CPEECC 2013). Self-selection becomes a challenge in recruiting participants with minority viewpoints. They can opt not to participate, despite being invited.

Advantage

The recruitment strategy, at its onset, did engage in a random recruitment process, meeting some of the criteria required to generalize the findings beyond deliberative participants. This recruitment process helped provide legitimacy in the eyes of City Administration and Council. The assumption was that the policy recommendations were on solid ground if all demographic groups were represented in the group of deliberating participants. In particular, council members talked about the importance of representing those employed in the energy sector as well as those not employed in the energy sector, and citizens who are skeptical of climate change (see chapter 1).

The recruitment strategy was successful in ensuring demographic diversity. The only major deviation was in terms of households with small children and those with high school education or less. In terms of the education bias, an Alberta study suggests that education does not affect beliefs about the existence of climate change, beliefs about the causes of climate change, or level of concern for climate change (Davidson and Haan 2012). However, as mentioned, public opinion polls may over-represent educated people and may not properly represent differences in views based on education (see prior discussion of Hall, Wilson, and Newman 2011). The deliberative participants were comparable to the census profile on the demographic variables that affect attitudes related to climate change, such as gender and age (Davidson and Haan 2012). This deliberative event was successful in establishing representativeness, but struggled with the inclusion of climate change deniers who were invited but chose not to participate. Further deliberative studies should experiment with approaches to more effectively engage and retain people holding minority viewpoints.

Case Study 3: Energy Efficiency Choices

This deliberation was organized by the Alberta Energy Efficiency Alliance in partnership with ABCD. Each of the deliberative groups met electronically for two hours sometime during November 2013. The recruitment of participants relied on Probit, using a process very similar to the process conducted for the Citizens' Panel on Edmonton's Energy and Climate Challenges, except that the goal was to engage a cross-section of Albertans. As mentioned, Probit conducts ongoing recruitment of citizens, via interactive voice response surveys (IVR), to participate in its online panel. Volunteers for this online panel were contacted by email to ask about interest in participating in a two-hour discussion about energy efficiency in Alberta. If they were interested, they were asked to type in their name and contact information as well as dates for which they would be available to participate.

Probit recruited 462 participants from their existing voluntary panel. However, many of these participants did not provide email addresses during this recruitment process, and so phone calls were made to all participants without email addresses asking them to confirm their interest in participating in the event. In the end, only 162 citizens participated in one of the series of two-hour meetings. In the Energy Efficiency Choices project, most participants were not contacted by an interviewer to confirm their participation. In contrast, in the Energy and Climate Challenges panel, all participants were contacted by an interviewer via phone to confirm their interest in participation, resulting in a higher participation rate. Of the four deliberative projects described, this project involved the least commitment of time from participants (two hours versus eight to forty-three hours for the other projects) but did require some technical skills in order to participate.

Despite using random recruitment, the process failed to recruit sufficient females, young people, people with less education, and people with young children. Approximately 29 per cent of the population is aged 18 to 34 years, whereas 2 per cent of those recruited were in this age group. Approximately 12 per cent of those with high school or less were recruited to participate, compared to 39 per cent of the Alberta population that has a high school education or less (Statistics Canada 2011). Finally, the recruitment process under-represented households with children in the home (Statistics Canada 2012) and included only 4 per cent Indigenous people, compared to the reported proportion of 6 per cent of the Alberta population (Statistics Canada 2011).

Disadvantage

Although this project involved the largest number of participants and some form of random recruitment, the representation was biased toward males and contained very few young people. Compared to the three offline deliberations, this deliberation had the poorest representation of young people in the deliberation (only 2 per cent were under the age of 30 years). Given the online form of the deliberation and the inclusion of cellphone lines, the under-representation of young people is surprising. This bias is important for the topic of climate change, as this group's future well-being may depend on effective climate change policies. Approximately 57 per cent of participants were male in this online deliberative exercise.

Advantage

The recruitment costs were approximately $10,000, which for a sample of 462 is cost-effective. However, as mentioned, this cost-effective approach depended on impersonal email correspondence without human contact via phone. This impersonal approach detrimentally affected the participation rate for the project. Many people expressed interest but failed to follow up on their commitment.

The cost-effective recruitment methods allow for the recruitment of a large sample, which is useful in examining nuances in attitude changes and policy preferences. In addition, the large sample and probability-based recruitment process may allow for the possibility of generalizing findings, after weighting to address under-representation and over-representation of key demographic groups. However, with the high dropout rate between recruitment and participation, the principle of randomness is seriously compromised. While 462 citizens expressed interest in participation, only 162 actually participated. Of the four deliberative projects, this was the highest dropout rate between recruitment and participation. This dropout rate could indicate non-response bias, in which the attitudes held by those who participated differ significantly from the attitudes of those who did not participate. This bias may undermine the ability of this group to represent the range of views about energy and climate issues in Alberta.

Case Study 4: Water in a Changing Climate Citizen Panel

This deliberative group was organized as a partnership between the Oldman Watershed Council and Alberta Climate Dialogue and involved citizens discussing the connection of water to climate change. The participants met for eight hours in February 2014 (see chapter 1). Participants were recruited through two methods. The first approach was to send invitations out through the Oldman Watershed Council's electronic mailing list (Alberta Climate Dialogue 2014). The second was through advertisements published in newspapers or at public meeting places, such as libraries and post offices (Alberta Climate Dialogue 2014). In either case, participants had to sign up on the ABCD website. As part of the sign-up procedures, participants were asked their name, contact information, occupation, length of residence in the Oldman Watershed area, whether they had specialized knowledge about climate change or water issues, their self-assessed knowledge level around climate change, and their views about whether climate change is happening and the sources of climate change (human, natural, combination). In total, sixty people signed up to participate in the deliberation held on February 22, 2014. Thirty-three people were selected from this list with the goal of ensuring a diversity of perspectives (Alberta Climate Dialogue 2014). Participants were offered $100 for participating (Alberta Climate Dialogue 2014).

The recruitment process produced slightly more women than men, an under-representation of young people, and under-representation of those with high school or less. The recruitment process was successful in recruiting people from First Nations communities. In terms of political ideology, the group was split evenly between right- and left-wing thinkers (Alberta Climate Dialogue 2014). In addition, the distribution of urban and rural dwellers matched the characteristics of the region (Alberta Climate Dialogue 2014).

Disadvantage

The biggest disadvantage of this form of recruitment is that it is not random. As such, it is unclear whether panelists' characteristics, as well as their views, are representative of the population. The recruitment method produced a list of participants who were more likely to be women, older, and better educated, but the misrepresentation on demographics was not greater than the bias observed for recruitment processes that included a random selection component (see Table 4.1). However, as mentioned, demographic representation does not mean attitudinal representation. Given the recruitment method, the greatest concern is that these participants were more politically aware than non-participants. The participants were largely recruited through the Oldman Watershed Council electronic mailing list and thus were a group already engaged with the political process. In addition, they may have had homogeneous views on climate change and water issues, because they were largely recruited from a single community organization.

Advantage

The greatest advantage of this form is the low cost. The recruitment through the electronic mailing list has minimal costs. The sign-up process required minimal work from a programmer. Finally, posting notices at public spaces in the community required minimal labour and printing costs. The targeted recruitment strategy was effective in representing groups that are under-represented in other processes (see Table 4.1). Approximately 9 per cent of participants were from First Nations communities. This group is typically left out of traditional policy-making processes. In terms of climate change, this group may be differentially affected by the impacts of climate change, which makes their participation critical (Blue 2012). With respect to First Nations, the recruitment strategy was able to achieve inclusiveness, but for other groups, representativeness is a key concern.

Conclusion

This chapter highlighted a number of important considerations and challenges in recruiting participants for deliberative projects, and described the recruitment processes followed for four public deliberations in which ABCD participated. Despite efforts to ensure representativeness, all four case studies were biased in terms of education. This challenge is consistent with other deliberative events, which also fail to reflect the educational composition of their geographic community (Farrar et al. 2010; Fournier et al. 2011; French and Laver 2009; Merkle 1996). The recruitment approach for the Citizens' Panel on Edmonton's Energy and Climate Challenges performed better than the other approaches because quotas were established around recruitment processes. Comparing four deliberative exercises demonstrates that larger and random samples do not better

Table 4.1. Demographic profile of recruited sample for the four deliberative projects

	Edmonton Food and Urban Agriculture n=58*	Edmonton's Energy and Climate Challenges n=66**	Energy Efficiency Choices n=462**	Water in a Changing Climate n=33*
Length of deliberation	2 days and 4 ½ days	6 days	2 hours	1 day
Honorarium	$150	$400	–	$100
Percentage of females	57%	52%	42%	55%
Percentage with children at home, under the age of 18 years	–	26%	23%	–
Indigenous	2%	3%	4%	9% from First Nations Communities
Age				
34 and under	29%	35%	18–29: 2%	15%
35–44	12%	12%	30–49: 26%	9%
45–54	20%	15%	50+: 72%	12%
55–64	19%	17%		33%
65+:	20%	21%		24%
Education				
High school or less	6%	29%	12%	6%
Some college, trade school, University, or completed diploma	24%	30%	34%	44%
University degree, certificate, or more	70%	41%	54%	50%

*Number represents actual participants rather than all those recruited.

**Number represents individuals recruited to deliberate, not the number of participants.

Sources: City of Edmonton 2012; Gwendolyn Blue, Email message to author, July 8, 2015; the author's analysis of data from Citizens' Panel on Edmonton's Energy and Climate Challenges; and Kristjana Loptson, email message to author, June 19, 2015.

represent the demographic characteristics of the population than smaller and non-random approaches. This was most evident in the Energy Efficiency Choices deliberative exercise. Furthermore, this chapter highlighted the trade-offs between representativeness and inclusiveness. The approaches used by the Water in a Changing Climate and the Food and Urban Agriculture Citizen Panel performed better at recruiting people who are typically excluded from the policy-making process, such as Indigenous people and visible minorities. However, these approaches fared the worst in terms of education bias. Only 6 per cent of participants had high school or less. In other words, inclusiveness came at the expense of representativeness. Fournier et al. (2011) suggest addressing education bias by examining whether education predicts differences in policy preferences.

In the case of deliberations about climate change, recruitment processes need to account for self-selection biases within random recruitment approaches, which may lead to the over-representation of people who are more interested and knowledgeable about the issues than others. The Citizens' Panel on Edmonton's Energy and Climate Challenges was well-positioned to identify and address attitudinal biases before the deliberative event, because this project included a large, random digital dialing survey of Edmontonians conducted prior to the event. This recruitment enabled the identification of bias in participation at the onset. This recruitment approach identified climate deniers and invited them to participate, but in the end, this group disengaged from the project. Perhaps this group could have been retained if they were over-sampled, providing a critical mass of participants with minority viewpoints (Bächtiger, Setälä, and Grönlund 2014). This approach of over-representation was used for First Nations residents in the Water in a Changing Climate, and was useful in ensuring inclusiveness.

While the discussion of representativeness focuses on education, gender, and age (Andersen and Hansen 2007; Farrar et al. 2009; Farrar et al. 2010; Fishkin et al. 2010; French and Laver 2009; Griffin et al. 2015; Hall, Wilson, and Newman 2011; Hansen and Andersen 2004; Hobson and Niemeyer 2011; Setälä, Grönlund and Herne 2010; Strandberg and Grönlund 2012), these four deliberative events identified another group that is challenging to recruit. Families with small children were difficult to engage in these deliberative events. Unfortunately, efforts to address this bias, such as offering free childcare to enable participation of families with small children, were ineffective in overcoming some biases in participation. Reducing the effort required to engage in the deliberative project was also ineffective in obtaining participation from this group. The Energy Efficiency Choices project required minimal effort to participate, but the project failed to engage participants who had small children in the home. Further research should experiment with alternative recruitment strategies to address participation biases and consider how these biases impact the policy recommendations proposed by deliberative groups. Finally, the literature on deliberative events should introduce standardized reporting approaches, like those offered in public opinion research (see https://www.aapor.org/), to enable comparisons across events about the number of people who were invited, the number of people who showed up, and the number of people who fully participated in the deliberative event (Karjalainen and Rapeli 2015). Different reporting approaches were used in the four deliberative events, reflecting differences in the broader literature's approach to reporting on participation.

References

Alberta Climate Dialogue. 2014. *Water in a Changing Climate*. Calgary: Alberta Climate Dialogue. http://www.albertaclimatedialogue.ca/final-report-water-in-a-changing-climate/final-report-water-in-a-changing-climate/.

Andersen, Vibeke Normann, and Kasper M. Hansen. 2007. "How Deliberation Makes Better Citizens: The Danish Deliberative Poll on the Euro." *European Journal of Political Research* 46(4): 531–56. doi:10.1111/j.1475-6765.2007.00699.x.

Bächtiger, André, Maija Setälä, and Kimmo Grönlund. 2014. "Towards a New Era of Deliberative Mini-Publics." In *Deliberative Mini-Publics: Involving Citizens in the Democratic Process*, edited by Kimmo Grönlund, André Bächtiger, and Maija Setälä, 224–46. Colchester, UK: European Consortium for Political Research Press.

Blue, Gwendolyn and Jennifer Medlock. 2014. "Public Engagement with Climate Change as Scientific Citizenship: A Case Study of World Wide Views on Global Warming." *Science as Culture* 23(4): 560–79. doi:10.1080/09505431.2014.917620.

Blue, Gwendolyn, Jennifer Medlock, and Edna Einsiedel. 2012. "Representativeness and the Politics of Inclusion: Insights from World Wide Views Canada." In *Citizen Participation in Global Environmental Governance*, edited by Richard Worthington, Mikko Rask, and Lammi Minna, 139–52. London: Earthscan.

City of Edmonton. 2012. *City-Wide Food and Urban Agriculture Strategy Report On Citizen Panel Process and Recommendations*. https://d1ok7k7myw g42z.cloudfront.net/assets/509af364dabe9d6d3b004a74/Citizen_Panel_Final_Report_September_2012.pdf.

CPEECC (Citizens' Panel on Edmonton's Energy and Climate Challenges). 2013. *Citizens' Panel on Edmonton's Energy and Climate Challenges Report*.

https://www.edmonton.ca/city_government/documents/PDF/CitizensPanel-EnergyClimateChallenge.pdf.

Davidson, Debra J., and Michael Haan. 2012. "Gender, Political Ideology, and Climate Change Beliefs in an Extractive Industry Community." *Population and Environment* 34(2): 217–234. doi:10.1007/s11111-011-0156-y.

Farrar, Cynthia, Donald P. Green, Jennifer E. Green, David W. Nickerson, and Steven Shewfelt. 2009. "Does Discussion Group Composition Affect Policy Preferences? Results from Three Randomized Experiments." *Political Psychology* 30(4): 615–47. doi:10.1111/j.1467-9221.2009.00717.x.

Farrar, Cynthia, James S. Fishkin, Donald P. Green, Christian List, Robert C. Luskin, and Elizabeth Levy Paluck. 2010. "Disaggregating Deliberation's Effects: An Experiment Within a Deliberative Poll." *British Journal of Political Science* 40(2): 333–47. doi:10.1017/s0007123409990433.

Fishkin, James S., Baogang He, Robert C. Luskin, and Alice Siu. 2010. "Deliberative Democracy in an Unlikely Place: Deliberative Polling in China." *British Journal of Political Science* 40(2): 435–48. doi:10.1017/s0007123409990330.

Fournier, Patrick, Henk van der Kolk, R. Kenneth Carty, André Blais, and Jonathan Rose. 2011. *When Citizens Decide: Lessons from Citizen Assemblies on Electoral Reform*. Oxford: Oxford University Press.

French, Damien, and Michael Laver. 2009. "Participation Bias, Durable Opinion Shifts and Sabotage Through Withdrawal in Citizens' Juries." *Political Studies* 57(2): 422–50. doi:10.1111/j.1467-9248.2009.00785.x.

Gastil, John. 2000. *By Popular Demand*. Berkeley: University of California Press.

Gastil, John, Katie Knobloch, and Meghan Kelly. 2012. Evaluating Deliberative Public Events and Projects. In *Democracy in Motion: Evaluating the Practice and Impact of Deliberative Civic Engagement*, edited by Tina Nabatchi, John Gastil, G.M. Weiksner, and Matt Leighninger, 205–30. Oxford: Oxford University Press.

Griffin, Jamie, Tarik Abdel-Monem, Alan Tomkins, Amanda Richardson, and Stacia Jorgenson. 2015. "Understanding Participant Representativeness in Deliberative Events: A Case Study Comparing Probability and Non-Probability Recruitment Strategies." *Journal of Public Deliberation* 11(1): 1–28.

Grönlund, Kimmo, Maija Setälä, and Kaisa Herne. 2010. "Deliberation and Civic Virtue: Lessons from a Citizen Deliberation Experiment." European Political Science Review 2(1): 95–117. doi:10.1017/s1755773909990245.

Hall, Troy, Patrick Wilson, and Jennie Newman. 2011. "Evaluating The Short- and Long-Term Effects of a Modified Deliberative Poll on Idahoans' Attitudes and Civic Engagement Related to Energy Options." *Journal of Public Deliberation* 7(1): 1–31. http://www.publicdeliberation.net/cgi/viewcontent.cgi?article=1136&context=jpd.

Hansen, Kasper M., and Vibeke Normann Andersen. 2004. "Deliberative Democracy and the Deliberative Poll on the Euro." *Scandinavian Political Studies* 27(3): 261–86. doi:10.1111/j.1467-9477.2004.00106.x.

Hobson, Kersty, and Simon Niemeyer. 2011. "Public Responses to Climate Change: The Role of Deliberation in Building Capacity for Adaptive Action." *Global Environmental Change* 21(3): 957–71. doi:10.1016/j.gloenvcha.2011.05.001.

James, Michael Rabinder. 2008. "Descriptive Representation in the British Columbia Citizens' Assembly." In *Designing Deliberative Democracy: The British Columbia Citizens' Assembly*, edited by Mark Warren and Hilary Pearse, 106–21. Cambridge: Cambridge University Press.

Karjalainen, Maija, and Lauri Rapeli. 2015. "Who Will Not Deliberate? Attrition in a Multi-Stage Citizen Deliberation Experiment." *Quality and Quantity* 49(1): 407–22. doi:10.1007/s11135-014-9993-y.

Loptson, Kristjana, and Shelley Boulianne. October 2013. "Trust in Municipal Government: The Case of the Citizens' Panel on Edmonton's Energy and Climate Challenges." Alberta Climate Dialogue. http://www.albertaclimatedialogue.ca/research-brief-trust-in-municipal-government/research-brief-trust-in-municipal-government//edit.

Lukensmeyer, Carolyn, and Steve Brigham. 2005. "Taking Democracy to Scale: Large Scale Interventions—For Citizens." *Journal of Applied Behavioral Science* 41(1): 47–60. doi:10.1177/0021886304272656.

Luskin, Robert C., James S. Fishkin, and Roger Jowell. 2002. "Considered Opinions: Deliberative Polling in Britain." *British Journal of Political Science* 32(3): 455–87. doi:10.1017/s0007123402000194.

Mao, Yuping, and Marco Adria. 2013. "Deciding Who Will Decide: Assessing Random Selection for Participants in Edmonton's Citizen Panel on Budget Priorities." *Canadian Public Administration* 56(4): 610–37.

Merkle, Daniel M. 1996. "Review: The National Issues Convention Deliberative Poll." *Public Opinion Quarterly* 60(4): 588–619. doi:10.1086/297775.

Neblo, Michael A., Kevin M. Esterling, Ryan P. Kennedy, David M.J. Lazer, and Anand E. Sokhey. 2010. "Who Wants To Deliberate – And Why?" *American Political Science Review* 104(3): 566–83. doi:10.1017/s0003055410000298.

Olcese, Cristiana, Clare Saunders, and Nikos Tzavidis. 2014. "In the Streets with a Degree: How Political Generations, Educational Attainment and Student Status Affect Engagement in Protest Politics." *International Sociology* 29(6): 525–45.

Ryfe, David Michael, and Brittany Stalburg. 2012. "The Participation and Recruitment Challenge." In *Democracy in Motion: Evaluating the Practice and Impact of Deliberative Civic Engagement*, edited by Tina Nabatchi, John Gastil, Matt Leighninger, and G. Michael Weiksner, 43–58. Oxford: Oxford University Press.

Setälä, Maija, Kimmo Grönlund, and Kaisa Herne. 2010. "Citizen Deliberation on Nuclear Power: A Comparison of Two Decision-Making Methods." *Political Studies* 58(4): 688–714. doi:10.1111/j.1467-9248.2010.00822.x

Statistics Canada. 2011. "NHS Focus on Geography Series – Edmonton." http://www12.statcan.gc.ca/nhs-enm/2011/as-sa/fogs-spg/Pages/FOG.cfm?lang=E&level=3&GeoCode=835.

———. 2012. "Edmonton, Alberta (Code 4811061) and Division No.11, Alberta (Code 4811) (Table). Census Profile. 2011 Census." http://www12.statcan.gc.ca/census-recensement/2011/dp-pd/prof/index.cfm?Lang=E.

———. 2015. "CANSIM - 115-0014 – Total Income for Adults with and without Disabilities, By Age Group and Sex, Canada, Provinces and Territories." http://www.statcan.gc.ca/daily-quotidien/150522/dq150522f-cansim-eng.htm.

Stewart, David W., Prem N. Shamdasani, and Dennis W. Rook. 2011. *Focus Groups: Theory and Practice.* Thousand Oaks, CA: Sage.

Strandberg, Kim, and Kimmo Grönlund. 2012. "Online Deliberation and its Outcome –Evidence from the Virtual Polity Deliberative Experiment." *Journal of Information Technology and Politics* 9(2): 167–84.

Weber, Elke U. 2010. "What Shapes Perceptions of Climate Change?" *Wiley Interdisciplinary Reviews: Climate Change* 1(3): 332–42. doi:10.1002/wcc.41.

Weber, Elke U., and Paul C. Stern. 2011. "Public Understanding of Climate Change in the United States." *American Psychologist* 66(4): 315–28. doi:10.1037/a0023253.

5

From Facts to Frames

Dominant and Alternative Meanings of Climate Change

Gwendolyn Blue

This chapter situates the specific deliberations developed by Alberta Climate Dialogue (ABCD) in the context of broader meanings, or frames, of climate change. The intent is to draw attention to the significance of framing for public deliberation with climate change and how deliberative framing can be applied to the organization of deliberations "on the ground."

I have been involved with ABCD as a researcher and an organizer of one of the deliberations—Water in a Changing Climate. My approach to public deliberation is informed by the interpretive social sciences. From my academic vantage point, existing forms of social power matter for the ways in which people understand, discuss, and come to decisions on environmental problems. In deliberation and elsewhere, people tell stories to get a handle on a complex and uncertain world. The language we use and the stories we tell do not innocently reflect reality. Rather, our stories actively shape the ways in which we perceive, understand, and act in the world. In turn, some groups have more power than others in presenting their accounts in the public sphere. I'm curious whether public deliberation, properly designed, can assist in bringing marginalized perspectives and values into conversation with dominant perspectives and values to foster reflection about and perhaps even reorientation of dominant beliefs.

My stance toward public deliberation is critical but not dismissive. My concern with formal face-to-face public deliberation, and particularly consensus-driven initiatives, is that they can all too easily reinforce rather than call into question dominant meanings and power relations. Research has repeatedly demonstrated that exclusionary practices and unequal power relations often structure these initiatives. In relation to science-based policy issues such as climate change, technical frames of reference typically trump other meanings that citizens might bring to the table.

My concern throughout the ABCD process, in workshops and in the delivery of initiatives, was that insufficient attention was paid to issue framing in the design and implementation of deliberation. While this concern applies to both the framing of deliberation and climate change, this discussion will focus primarily on the latter. For the most part, climate change was approached as a technical problem of mitigation (i.e., efforts to reduce greenhouse gases) and of energy. The implications of considering and grappling with multiple frames of climate change were not widely explored by this research-practitioner group.

This chapter tells the story of my efforts to expand the frames of climate change in ABCD. Given space restraints, this story is necessarily partial and limited. To begin, I discuss the significance of framing for public deliberation. Next, I outline two frames of climate change: a mainstream approach that emphasizes mitigation of greenhouse gas (GHG) emissions through primarily market-based or technological measures; and an approach that emphasizes other policy dimensions such as adaptation (efforts to address and cope with weather and climate extremes). I illustrate how alternative frames of climate change were taken up in a one-day deliberation (Water in a Changing Climate) and discuss the implications of expanding the frames of climate change for deliberative purposes. I conclude with recommendations for future practice.

Shared Meaning: Framing Issues for Public Deliberation

In deliberative settings, citizens are typically asked to reflect on their own values as well as the values of others. The focus tends to be on an individual's ideas, interests, and values and how these evolve through interaction and deliberation. The underlying assumption is that meaning is individual. Interpretive research tells us, however, that an individual's values and beliefs are inherently social. The stories that we tell and the meanings that we give to particular issues are strongly influenced by shared ways of making sense of the world (Dryzek 2013). These shared meanings are issue frames that help us make sense of the world by directing our attention to certain aspects of reality and not others (Entmann 1993). Framing refers to the ways in which problems are defined, causes are diagnosed, and remedies are suggested. Framing is an inherent and normal part of communication. Since we cannot avoid framing, the best we can do is to acknowledge its effects and manage its consequences.

Research suggests that when people are generally not well informed about an issue, the ways in which information is presented heavily influences responses. For instance, a significant experiment in cognitive psychology sought to determine if framing played a role in informing consumer preferences and risk judgments (Tversky and Kahneman 1986). At the time of this experiment, it was widely believed that preferences, opinions, and judgments were largely stable and individually determined. This study demonstrated, to the contrary, that if people are presented with an uncertain situation, their preferences and attitudes change depending on the ways in which the information at hand is presented. This research raises an important question: Who or what controls the opinions, values, and "voice" of citizens if the information with which they are provided plays such a powerful role in how they think about policy issues?

In some cases, certain frames are selected to guide and control the conversation. A common political strategy is to present messages from the perspective of a narrow frame to get people to respond in a predetermined way. This approach, known as framing to persuade, involves advancing or favouring one frame over others to bring people on side. Framing to persuade is common in environmental communication, and is an increasingly popular strategy in climate communication (Lakoff 2010).

When issues are framed for persuasion purposes, the arguments and courses of action are established in advance and the focus is largely instrumental. By contrast, framing for deliberation seeks to present and clarify different ways of approaching an issue to help people weigh appropriate courses of action (Friedman 2007; Calvert and Warren 2014). The goal is largely substantive, which is to say that this type of framing is intended to help people come up with potentially innovative solutions that they wouldn't have reached prior to engaging with one another. Presented with a diverse range of frames from the outset, people are better equipped to make sense of competing values and arguments and not be "boxed in" by a singular approach.

Formalized public engagement initiatives can limit policy options by offering a small range of options to participants from the outset (Pallett and Chilvers 2013) or by enabling framing effects wherein dominant frames shut down other possibilities and lead to a premature closing down of policy options (Calvert and Warren 2014). Although multiple and conflicting issue frames are present in any policy discussion, dominant frames can limit the discussion because they appear to be taken for granted, as the way things are, rather than as contestable policy options. Dominant frames can lead to framing effects which include groupthink, premature closure of options, or forced consensus.

Framing effects can be avoided or mitigated in formal public deliberation if the process accounts for different frames in the design, implementation, and delivery of initiatives. For instance, organizations such as the National Issues Forums (NIF) use a choice work frame that presents multiple perspectives on policy issues (Friedman 2007; Kadlec and Friedman 2007). Issue guides provide participants with an overview of dominant and alternative frames of a policy issue, and the values that are contained therein. By presenting multiple perspectives on an issue, the NIF enables a more deliberative approach to political engagement than would be possible if only one issue frame were presented. This approach can also facilitate a deeper reflection among deliberative practitioners, scientific experts, bureaucrats, and policy makers on their own existing assumptions and values (Pallett and Chilvers 2013).

Framing Climate Change for Deliberation

Although climate change is a complex issue with many different policy frames, a dominant frame circulates among policy makers, scientists, civil society groups, and citizens. It is the frame most commonly encountered in the media, in social activism campaigns, and in government policy. This frame provides a readily recognizable story in which the problem of human-caused climate change is connected specifically to GHG emissions, primarily carbon dioxide, and warrants responses such as technological, market-based, or behavioural change. The dominant frame of climate change privileges the knowledge of scientists, engineers, and other experts such as economists (Hajer and Versteeg 2005). It emphasizes incremental reforms rather than radical changes to existing systems. An increasing number of activists and academics highlight the limitations of this approach for addressing climate change, and the need to provide alternative frames for public consideration (Klein 2014; Hulme, 2009; Dawson 2010).

To understand how this dominant frame emerged, a brief history is in order. In the 1970s and 1980s, networks of scientists and government experts played a key role in putting climate change on international and national political agendas (Bulkeley et al. 2014). In 1988, the World Meteorological Organization and United Nations Environmental Programme formed the Intergovernmental Panel on Climate Change (IPCC). This organization was initially charged with providing an assessment of relevant research to direct policy. After its first assessment process, the IPCC presented its summary reports as providing scientific information that informs but does not give direction on the actions policy makers should take. The first report of the IPCC (released in 1990) became the basis for the United Nations Framework Convention on Climate Change (FCCC) at the Rio de Janeiro Earth Summit in 1992. By the end of this conference, the FCCC was signed by 154 states and entered into force in March 1994. The FCCC is the legal instrument of the global climate regime. Its mandate is to work toward stabilizing greenhouse gas emissions to prevent dangerous climate change. Negotiations under the FCCC led to the Kyoto Protocol in 1997 as well as the Copenhagen Accord in 2009.

The mandate of the FCCC emphasizes mitigation of dangerous climate change through the reduction of GHG emissions, and this mandate has had a strong influence on the ways in which climate change is approached at national and regional levels. One of the reasons that mitigation is highlighted in the FCCC is that lessons were drawn from previous global environmental policy responses, such as ozone depletion, in which mitigation measures were successful (Pielke 1998). Another explanation is that the emphasis on mitigation was in keeping with the interests of wealthy industrialized countries and not with those regions in which the effects of climate change were already being experienced. As Okereke explains, the focus on mitigation in the FCCC can be attributed to concerns by wealthier states "that an emphasis on adaptation would greatly provoke questions of responsibility, liability and the polluter pays principle—all of which they were anxious to avoid during the negotiation process" (Okereke 2008, 105–6).

The IPCC is based on a linear model of scientific expertise, which is to say that the interactions between science and policy are unidirectional, with the assumption that science informs policy by speaking truth to power. At first glance, the basic logic of such an approach is sensible. The relevant facts about climate change should be established before deciding what policies to implement. In following a linear model of expertise, however, politically relevant questions are often framed in a way that detaches expertise from its political and cultural contexts. The types of policy measures that follow tend also to be highly technical in nature. Propelled by a belief in the neutrality of science, the IPCC typically avoids addressing value-based decisions and openly advocating or rejecting policy options. The linear model of expertise also tends to stifle discussions about alternative policy options, some of which might radically challenge existing status quo practices. As such, climate change tends to be framed as a relatively tame problem that can be solved by technological solutions or market-based mechanisms that keep broader political and economic systems in place.

To date, there has been little public discussion of the assumptions and value commitments embedded within this dominant frame of climate change. This can be attributed in large part to the complexity of climate science and the highly specialized bodies of expert scientific knowledge through which it has been understood (Demeritt 2001, 2006).

ABCD concentrated most of its efforts on this dominant frame of climate change, namely, by examining the intersections between climate and energy, with an attendant focus on mitigation. An alternative approach is to consider climate change as a nexus that connects water, food, and energy (O'Riordan and Sandford 2015). From this perspective, adaptation is also a viable response. While there is no question that mitigation strategies are necessary to address climate change, many argue that we have passed the point at which mitigation measures alone are sufficient (Craig 2010). Mitigation efforts at global and national levels have proven largely ineffective as carbon emissions continue to rise. The global warming experienced so far is already driving climatic change in regions around the world, and this change is expected to accelerate in the future. Moreover, the changes that will happen will have dramatic and, in many cases, unpredictable consequences.

As a matter of international law, climate change adaptation is a key component of the FCCC. Yet compared to mitigation, adaptation has not received the same level of attention from policy makers, civil society groups, the media, or the general public. Part of this is due to the historical marginalization of adaptation in global discussions of climate change. Some reasons for lack of attention to adaptation are that it is associated with passive acceptance or fatalism, that it will take attention away from mitigation, that it is not in the interests of northern industrialized regions, and that its inherently local characteristics make it difficult to distinguish regional or local climate impacts from global circulation models (Pielke 1998; Rayner and Malone 1998). These assumptions are changing, however, and adaptation is increasingly receiving more attention at global and local levels, particularly in regions where the effects of climate change are already being experienced.

In dominant frames of climate change, adaptation tends to be seen as a tag-on to mitigation, and risk-based, technical approaches are common (Khan and Roberts 2013). Alternative approaches to both adaptation and mitigation highlight their social dimensions by foregrounding inequality and the need for justice to address the causes and consequences of climate change (Hackman, Moser, and St. Clair 2014).

Alternative Frames for Climate Change: A Pilot Project

The one-day event, Water in a Changing Climate, was an attempt to bring an alternative frame of climate change into ABCD deliberations. The intent of the panel was to expand the frames of climate change in the design and execution of the deliberation and to widen the geographical reach of ABCD's face-to-face deliberations to include rural populations. Our initial hope was that we could help participants transcend the common assumption that climate change is a distant risk in space and time. Linking climate change with water helped us focus on the tangible dimension of global environmental change. Some of the most pronounced and harmful impacts of global climate change are experienced through water. Communities in Alberta currently struggle with water-related challenges, including droughts, flooding, water pollution, and depletion of groundwater. These issues are compounded by global climate change (Henderson and Sauchyn 2008).

It is important to take note of some important parameters of this panel from the outset. First, limited resources meant that we were only able to deliver this as a one-off initiative. Since we did not have the opportunity to test this design or to build on our learnings, this event should be understood as a pilot project and not as a best-case example. In turn, our partner, the Oldman Watershed Council, had limited time and resources to devote to the event. These constraints reflect real world conditions in which cash-strapped institutions and time-strapped individuals are tasked with designing and delivering public engagement. Understanding these constraints from the outset can hopefully redirect the tendency of those who seek to place blame for any shortcomings on individuals or institutions. Second, unlike the other deliberative initiatives ABCD members were involved in, this deliberation did not have a pressing policy framework or decision. Indeed, the absence of an existing policy framework provided us with flexibility with respect to how we framed climate change.

A core partner for this project was the Oldman Watershed Council (owc), a not-for-profit organization in southern Alberta, mandated by the provincial government to provide guidance around the management and health of the Oldman Basin. In Alberta, and in Canada more broadly, a common approach to environmental governance of water is one that focuses on a specific area of land that drains water to a shared destination. A "watershed approach," as it is called, is central to the Alberta government's Water for Life strategy as well as the province's emergent land use strategy. As part of the Water for Life strategy, Watershed Planning and Advisory Councils (wpacs) have been established in major watershed basins. These councils are multi-stakeholder, non-profit organizations that assess watershed conditions and develop management plans. The owc's sixteen-member board of directors has representation from rural municipalities, academia, irrigation districts, environmental ngos, the agricultural industry, First Nations communities, and the federal government.

Citizen engagement is central to the mandate for wpacs (Alberta Water Portal 2017), although the owc faces several challenges with respect to public engagement. In the past, public engagement has defaulted to town hall–style public meetings that tend to attract "usual suspects" and "worried spies" rather than a broader diverse constituency (Frank 2013). The owc expressed a desire to learn more about deliberation as a possible avenue for broadening their approach to citizen engagement.

Prior to the panel, the project manager, Erin Navid, and I developed a discussion guide to provide participants with an overview of three different frames of climate change. These frames were drawn from existing literature on climate change and were presented as a starting point for discussion. The frames were described as follows:

> *Climate change as a problem that can be solved*: Climate change is a
> problem that humans can and should solve through reducing green-
> house gas emissions. Dangerous climate change can be prevented
> through technology, markets, or behavioural changes.

> *Climate change as an issue of justice*: Approaching climate change as
> an issue of justice means thinking about the ways in which people
> and other living creatures are vulnerable to weather and climate.
> This perspective addresses the human and cultural, as well as
> physical components of climate change. Reducing greenhouse gases
> is important but not sufficient to address climate change. Building

resilience to weather and climatic changes, confronting social inequality, and addressing stewardship for the natural world are also significant.

Climate change as a force of nature: This perspective emphasizes forces that influence the climate that are outside of human control. People who hold this perspective tend to believe that there is little we can do to prevent climate change. These people also tend to be skeptical of information that suggests otherwise.

The first perspective represents a dominant approach to climate change, although it was not described as such. The second represents a social justice frame that includes issues of adaptation. The third represents what is typically called the "denier" position. These frames represent different policy directions and assumptions but are by no means the only ways in which climate change could be approached. For example, mitigation can also encompass justice issues, and adaptation can be understood from a technical perspective. These frames are also not discrete in that people can hold several of these beliefs at the same time.

These three frames were used as a warm-up exercise wherein participants were asked to align themselves in the room based on where they felt they fit within these different approaches. Most participants aligned with the first frame. The remainder aligned with the social justice frame. Only two participants aligned themselves with the third frame, which views climate change as a force of nature.

The morning session was then dedicated to a discussion of the collective concerns about water and climate change in the region. The lead facilitator and designer of the deliberation, Jacquie Dale, with assistance from the table facilitators, categorized these concerns into the following themes: land use pressures; environment and public health; extreme weather events; governance; and social justice and responsibility. Synthesizing the diverse perspectives that emerged from the morning session into discrete categories was a challenging task.

In the afternoon session, participants were instructed to form groups based on which themes they found the most appealing. They were also asked to provide direction for the owc in moving forward. The recommendations emerging from the afternoon sessions included:

- Encourage regulation at a local level.
- Provide more education and information about how to deal with extreme weather events.

- Standardize emergency response plans, with timely and easy-to-access information
- Explore incentives to promote conservation and effective use of water.
- Foster individual stewardship for development of the common good.
- Support sustainable farming and agriculture, particularly in urban contexts.

Overall, the panel highlighted the importance of education, information, and communication as well as the significance and challenge of fostering collective responsibility for environmental protection. The following values were identified as central: healthy environment, public safety, stewardship, and collective responsibility.

The initial concern with social justice did not make it directly into the final recommendations, for reasons that are largely unknown. Although speculative, a partial explanation might be that participants were instructed by the owc to provide practical and tangible advice moving forward and social justice is difficult to fit within this directive. Differing facilitation skills across the individual table facilitators may also be part of the challenge of drawing out the implications of this frame for action. Devoting more time overall to the social justice dimensions of climate change, through facilitator training, the background document, and invited speakers, might have supported participants in relating the justice frame to specific recommendations for action.

While this citizen panel enabled a broader framing beyond the dominant frame of climate change, it had significant limitations. As mentioned previously, the alternative framing was possible because the initiative was not tethered to an existing policy conversation. While this offered freedom to explore alternative frames, it lacked policy relevancy. The one-off nature of the event meant that there was also no opportunity to build on the results to further collective learning, for instance, about the design and the frames that were deployed. It would have been useful to include the owc board more directly in the discussion to understand range of values that they bring to the policy and management strategies they put forth.

Another significant shortcoming of this initiative is that we did not address Indigenous world views of human–environment relations in our initial framings. It is well recognized that public deliberation initiatives—particularly those that are consensus oriented—can play a powerful role in silencing marginalized perspectives (Young 2000). This is not to say that efforts were not made to bring

Indigenous perspectives into the event. For instance, an image created by one of the panel participants, who received an additional honorarium for its development and use, was central in the event's communications. The image conveyed a message about the deep significance of water and culture from a local Blackfoot perspective. According to its illustrator, the image portrays future generations who must learn the importance of water and the environment. The background is in the shape of a hide which is typically used to document the histories and stories of Blackfoot peoples. In addition to this image, the panel began with a prayer by the designer of this image, who acknowledged the dual settler and First Nations governance of the area. These small gestures for inclusion could have been strengthened by drawing on Indigenous expertise in framing climate change and weather–related challenges from the outset. Providing a range of expertise on Indigenous world views, in addition to the views of climate scientists, would have fostered a more inclusive process that broadened even further the types of issues presented to participants. More critical attention to structural issues of inequality is also warranted, not only in terms of the design and delivery of deliberation but also in terms of the frames and assumptions about public deliberation that were deployed by ABCD. An absence of attention to issues of power means that public deliberation in practice can serve to reinforce rather than challenge existing social patterns of inequality.

Conclusion

The purpose of this chapter has been to draw attention to the various ways in which climate change can be framed and the implications for public deliberation. Frames are central to the ways in which we understand policy problems in terms of their causes and potential solutions. In deliberative settings, people must have a genuine opportunity to discuss, propose, and promote alternative frames. If deliberation is structured around a dominant or singular frame, people can feel disenfranchised and are more likely to disengage from the deliberative process. Although this explanation is speculative, a limited framing of climate change and of public deliberation might be one reason why this research process struggled to engage and retain a broader constituency of interested participants. Anecdotally, I know of several people who left this research collaboration because they felt that their concerns and perspectives on climate change and democratic engagement were not represented or valued by the group.

Opening deliberation to a range of frames is a significant part of the democratic process for substantive reasons. Frames are continuously reconstructed and reimagined as new participants, novel perspectives, observations, values, and world views are drawn into the mix. Different ways of knowing and experiencing environmental problems emerge from divergent social locations and experiences. This diversity is not a problem to be overcome but is a generative part of public responses to climate policy directives.

The core recommendation of this chapter is that, at minimum, organizers responsible for convening public deliberation on climate change make efforts to identify the range of available frames of climate change and to acknowledge the frames that they hold personally. This demands familiarity with not only the science of climate change but also the cultural and historical dimensions of this pressing policy issue. The interpretive social sciences and humanities are important allies in this regard (Blue 2015). Those interested in interpretive approaches to public deliberation with climate change can explore my work further (see, for instance, Blue 2015; Blue and Dale 2016; Blue 2017) and to contact me to learn about current unpublished initiatives.

While the temptation to gloss over issue framing is considerable, doing so presents significant problems. Considerable differences exist in the ways in which climate change is approached and interpreted, not only between expert and lay communities but also within academic communities. Providing participants with limited policy frames circumscribes the democratic potential of public deliberation. Practical constraints should not be used as a justification for avoiding the difficult task of grappling with the implications of framing and its effects on public deliberation with climate change.

References

Alberta Water Portal. 2017. Accessed March 21. http://albertawater.com/work/alberta-water-community/wpacs.

Blue, Gwendolyn. 2015. "Framing Climate Change for Public Deliberation: What Role for the Interpretive Social Sciences and Humanities?" *Journal of Environmental Policy and Planning* 18(1): 67–84.

———.2017. "Participatory and Deliberative Approaches to Climate Change." *Oxford Research Encyclopedia of Climate Change.* Accessed November 9, 2017, from http://climatescience.oxfordre.com/view/10.1093/acrefore/9780190228620.001.0001/acrefore-9780190228620-e-397.

Blue, Gwendolyn, and Jacquie Dale. 2016. "Framing and Power in Public Deliberation with Climate Change: Critical Reflections on the Role of Deliberative Practitioners." *Journal of Public Deliberation*, 12(1): art. 2. http://www.publicdeliberation.net/jpd/vol12/iss1/art2/.

Bulkeley, Harriet, Liliana Andonova, Michele Betsill, Daniel Compagnon, Thomas Hale, Matthew Hoffman, Peter Newell, Matthew Peterson, Charles Roger, and Stacy Vandeveer. 2014. *Transnational Climate Change Governance*. New York: Cambridge University Press.

Calvert, Aubre, and Mark Warren. 2014. "Deliberative Democracy and Framing Effects: Why Frames are a Problem and How Deliberative Mini-publics Might Overcome Them." In *Deliberative Mini-publics: Involving Citizens in the Deliberative Process*, edited by K. Grönlund, A Bächtiger, and M Setälä, 203–24. Colchester, UK: ECPR Press.

Craig, Robin. 2010. "'Stationarity is Dead'– Long Live Transformation: Five Priniciples for Climate Change Law." *Harvard Environmental Law Review* 34: 9–15.

Dawson, Ashley. 2010. "Climate Justice: The Emerging Movement against Green Capitalism." *South Atlantic Quarterly* 109(2): 313–38.

Demeritt, David. 2001. "The Construction of Global Warming and the Politics of Science." *Annals of the Association of American Geographers* 91(2): 307–37.

———. 2006. "Science Studies, Climate Change and the Prospects for Constructivist Critique." Economy and Society 35: 453–79.

Dryzek, John. 2013. *The Politics of the Earth: Environmental Discourses*. 3rd ed. Oxford: Oxford University Press.

Entmann, Robert. 1993. "Framing: Toward the Clarification of a Fractured Paradigm." *Journal of Communication* 43: 51–58.

Frank, Shannon. 2013. Personal Communication with Author.

Friedman, Will. 2007. "Reframing Framing." Occasional Working Paper. Public Agenda Center for Advances in Public Engagement.

Hackman, Heide, Susanne Moser, and Asuncion Lera St. Clair. 2014. "The Social at the Heart of Global Environmental Change." *Nature Climate Change* 4: 653–55.

Hajer, Maarten, and Wytske Versteeg. 2005. "A Decade of Discourse Analysis of Environmental Politics: Achievements, Challenges, Perspectives." *Journal of Environmental Policy and Planning* 7(3): 175–84.

Henderson, Norman, and David Sauchyn. 2008. *Climate Change Impacts on Canada's Prairie Provinces: A Summary of Our State of Knowledge*. Regina: Prairie Adaptation Research Collaborative.

Hulme, Mike. 2009. *Why we Disagree about Climate Change: Understanding Controversy, Inaction and Opportunity*. Cambridge: Cambridge University Press.

Kadlec, Amy, and Will Friedman. 2007. "Deliberative Democracy and the Problem of Power." *Journal of Public Deliberation* 3: 1–26.

Khan, Mizan, and Timmons Roberts. 2013. "Adaptation and International Climate Policy." *WIRES Climate Change* 4(3): 171–89.

Klein, Naomi. 2014. *This Changes Everything: Capitalism Versus the Climate*. New York: Simon and Schuster.

Lakoff, George. 2010. "Why It Matters How We Frame the Environment." *Environmental Communication: A Journal of Nature and Culture* 4: 70–81.

Okereke, Chukwumerije. 2008. *Global Justice and Neoliberal Environmental Governance*. London: Routledge.

O'Riordan, Jon, and Robert Sandford. 2015. *The Climate Nexus: Water, Food, Energy and Biodiversity in a Changing World*. Calgary: Rocky Mountain Books.

Pallett, Helen, and Jason Chilvers. 2013. "A Decade of Learning about Publics, Participation, and Climate Change: Institutionalising Reflexivity?" *Environment and Planning A* 45(5): 1162–83.

Pielke, Roger, Jr. 1998. "Rethinking the Role of Adaptation in Climate Policy." *Global Environmental Change* 8(2): 159–70.

Rayner, Steve, and Elisabeth Malone. 1998. "Ten Suggestions for Policy Makers." In *Human Choice and Climate Change, An International Assessment, Volume 4: What Have We Learned?*, edited by Steve Rayner and Elisabeth Malone, 3–32. Columbus, OH: Battelle Press.

Tversky, Amos, and Daniel Kahneman. 1986. "Rational Choice and the Framing of Decisions." *Journal of Business* 59: 251–72.

Young, Iris Marion. 2000. *Inclusion and Democracy*. Oxford: Oxford University Press.

6

Collaborating on Deliberative Democracy

David Kahane and Lorelei L. Hanson

Major deliberation projects—whether mini-publics or more complex initiatives—typically involve collaboration between deliberation professionals or organizations and some combination of academics, representatives of governments, NGOs, businesses, and others. Alberta Climate Dialogue was an unusually sustained research collaboration around deliberative democracy, and involved several further collaborations around deliberation projects. In this chapter, we unpack our learning about how to structure collaboration to support decision making and collective learning in complex deliberation projects.

Collaboration is "a mutually beneficial relationship between two or more parties to achieve common goals by sharing responsibility, authority and accountability for achieving results" (Chrislip 2002). When disparately positioned people collaborate, they share knowledge and learn collectively; engaging in cooperation and coordination, they create a mutual vision and joint strategies (Chrislip 2002).

Every context and community differs: no one model of collaboration works in all situations, yet many collaborative scholars agree on several fundamental elements required to address tough social problems. First, collaboration should include a range of stakeholders representing diverse interests, organizations, or perspectives on the issue of concern (Cestero 1999; Chrislip 2002; Mattessich and Monsey 1992): "If competing values and differing positions mark public problems, the work of defining problems and solutions must be done by the people who hold these values and positions" (Chrislip 2002, 45). Second, the group must identify attainable goals and objectives that all participants can agree are worthwhile (Mattessich and Monsey 1992; Williams and Ellefson 1996). Third, collaborative process should be well-designed, so that as the group works toward consensus there is a focus on negotiation and reflection (Keen and Mahanty 2006). Skilled facilitation helps stakeholders work together constructively, and content experts contribute appropriate knowledge and experience (Chrislip 2002; Williams and Ellefson 1996). Fourth, sustained commitment to the process, and shared power within it, enable a sense of shared responsibility to solve problems (Chrislip 2002; Weber 2000). Finally, where collaboration involves government, the collaborative process should be endorsed by key officials (Lampe and Kaplan 1999).

Collaboration that integrates these five fundamental elements supports the development of strategies needed to address wicked issues, which have no clear right answers, involve interconnected systems, and require adaptive responses (see the introduction). Collaboration that supports citizen engagement and deliberation can increase society's capacity to address wicked issues like climate change. Public deliberation can identify common purposes and develop collaborative relationships among citizens, public officials, and stakeholders; it increases players' abilities to respect and listen to one another's opinions, so that competing perspectives are aired and considered before decisions are made (Schusler, Decker and Pfeffer 2003). It also directs citizens and leaders to make value judgments and trade-offs among competing problem definitions and solutions.

Most public deliberation projects are complex collaborations, involving champions who initiate a project, funders, participation professionals, civic organizations, and often government officials and academics. Such collaborations can marshal resources and capacity for projects, increase the legitimacy of processes relative to those convened by a single player, and enhance the quality of deliberations by bringing multiple perspectives and networks to bear on design. Yet this diversity of collaborators also means working through variable understandings of the objectives of a deliberation, what constitutes effective social change, how best to approach difficult issues, and the risks involved. Collaborators bring divergent knowledge, best practices, ways of navigating political bureaucracies, and public communication styles. Conflicting power dynamics and different norms can produce tensions among actors, and individual actors may experience conflicting allegiances and identifications (Newman et al. 2004).

Each of the four deliberations that Alberta Climate Dialogue (ABCD) convened or participated in involved its own complex collaborations and context, and was enabled or constrained in distinctive ways, including the degree to which each used deliberation to engage citizens on climate change. ABCD partners included the Centre for Public Involvement (a research and practice organization co-funded by the University of Alberta and the City of Edmonton), a municipal government (the City of Edmonton), a non-government organization (the Alberta Energy Efficiency Alliance) and a para-governmental organization (the Oldman Watershed Council). The authors were engaged in these collaborations in different ways. David Kahane was the Project Director of ABCD, and intimately involved in the collaboration and resulting deliberation with the City of Edmonton, whereas Lorelei Hanson acted as a researcher on three of the projects. Our analysis of the collaborations between ABCD and these partners builds on our own recollections of group conversations and debriefs in meetings over the life of the project, semi-structured, taped interviews conducted with the partners and members of ABCD, and observational notes taken during the deliberations. Some of the interviews were conducted during the development of, or closely following, a deliberation and others were undertaken several years after the projects were complete.

Using the key collaborative elements described above as frames of analysis, we provide a short case study of each deliberation to explore the challenges, tensions, strengths, and opportunities encountered during the collaborations central to each of these projects. In looking at each case we trace:

- initial relationships and relations of trust that gave rise to the collaboration and how these influenced the development of goals and objectives of the deliberation;

- reasons for collaborating, including the amount of emphasis the outside partner placed on the outcome of the deliberation, and the associated perceived risks;

- the trajectory of development of the collaboration, particularly how conflicts about the design of the deliberation process or research manifested and were managed, and the role of outside decision makers in the process;

- the duration of the project, the time invested by collaborators, the burdens of contribution, and how these factors impacted trust and shared responsibility; and

- the forms of learning that emerged from the project among the convenors and citizen deliberators.

We will start, however, with an overview of ABCD as a collaboration.

Collaboration and Alberta Climate Dialogue

Alberta Climate Dialogue (ABCD) was a learning project, aiming to advance both the field of deliberative democracy and the quality of climate change responses through research on climate deliberations convened with organizations in Alberta. Over the seven years ABCD was active, it connected with dozens of people and organizations, some new to deliberative democracy and others expert in the field. Many members of ABCD took on their roles in addition to their job responsibilities. Academics in the project came from an institutional culture that emphasizes academic research for publication and does little to recognize or reward community-engaged scholarship, much less collaboration for political action. At the same time, many academics were concerned about climate change, and saw their involvement in ABCD as a political commitment. Deliberation professionals in the project also placed a strong emphasis on their own learning and on advancing the field, often regarding this learning as requiring different methodologies and forms of dissemination than those favoured by academics.

The practitioners most involved in the project contributed a lot of time pro bono, though several also received remuneration for consulting on particular deliberation projects. Participants in the project from NGOs and businesses were more interested in influence and action than in research; most remained detached from aspects of the project focused on learning, and some dropped out of the project because they saw its learning focus as not fitting with their motivation for involvement. A group of about ten researchers and deliberation professionals played a pivotal role in ABCD, working on the project in its early stages, serving on its steering committee, co-designing major workshops, leading research activities, and maintaining and sharing an understanding of the value of public deliberation in addressing tough political problems.

The federal research grant ABCD received provided much-needed funds and other institutional support, and placed expectations on ABCD to produce robust research and scholarship within its five-year mandate based on actual deliberations held with partners. This created a lot of pressure. For the first eighteen months or so of the project we were not sure that we would find partners for deliberations; ABCD thus became willing to conform to the needs of deliberation partners even when partners pushed against some of our judgments and principles around best process design, social learning, the link between deliberation and decision making, or the public profile of deliberations. ABCD's international team of engagement researchers and practitioners were leaders in theories and practices of innovative citizen involvement, yet we were selling approaches that governments and other organizations in Alberta were not actively seeking.

Case Study 1: The City of Edmonton's Food and Urban Agriculture Strategy

ABCD's first collaboration was not focused on conducting a deliberation but on partnering around the research component of a deliberation process already under way. In 2012, CPI was conducting a City-Wide Food and Urban Agriculture Citizen Panel with the City of Edmonton as part of developing a food and urban agriculture strategy. There were loose connections between members of ABCD and CPI, and two ABCD researchers were active in urban food politics and familiar with the political process under way; this led to ABCD members collaborating around research dimensions of the project.

The purpose of the citizen deliberation was to "develop recommendations that would be given to the City of Edmonton that would directly be included and inform the development of their urban food strategy" (Cavanagh 2015); the deliberation linked to climate change only indirectly. But ABCD was eager to begin research in the absence of its own deliberation projects, and CPI, an organization co-funded by the City of Edmonton and the University of Alberta to do research to support innovative public involvement, wanted more hands in developing a research strategy and more tools to bring to the deliberation project. ABCD members were interested in the research as the basis for comparative case studies, while CPI's realization of its mandate depended on producing strong research in the context of a deliberation project that was being developed on a timeline of only a few months in a "highly politicized context" (Cavanagh 2015).

When ABCD was first introduced to the food and urban agriculture deliberation, CPI already had a survey-based research strategy mapped out, which allowed little time to revisit major components of their project. CPI, a small organization, had only one full-time researcher devoted to collecting data for the project. Four ABCD researchers became involved in the research: two helped CPI design its citizen surveys and two undertook observation of the deliberations, conducted semi-structured interviews with six of the citizen deliberators, and consulted with CPI's researcher on the citizen survey results. However, there was friction between members of the ABCD research team—who were interested in exploring alternative approaches to this public deliberation, introducing more discussion of climate change into the framing, and examining both the challenges and strengths of the deliberation—and CPI researchers, who welcomed help on their existing track of research and implementing a specific approach to the deliberation. It should be noted that the players involved here were mostly academics: while much of this chapter examines challenges that arise in collaborations across boundaries of sectors and professions and organizations, here we see friction that can arise among those working in universities. This friction was not explicitly dealt with among the academics in this case; rather, tensions played out around the limited influence ABCD researchers were able to have on research methodology, and their restricted access to the research data.

The collaboration between ABCD and CPI on the City-Wide Food and Urban Agriculture Citizen Panel was thus very limited. The deliberation process appeared to ABCD's researchers to be robust and well facilitated. Shared learning was constrained by the lack of both trust and shared responsibility between ABCD and CPI, and the politically sensitive context for the deliberation. As a result, there were limited opportunities for ABCD to use CPI's experiences to inform their deliberative processes, and for CPI to consider the feedback and analysis provided by ABCD researchers on the Citizens' Panel.

Case Study 2: The City of Edmonton's Citizens' Panel on Edmonton's Energy and Climate Challenges

ABCD's second collaboration was with the City of Edmonton's Office of the Environment, and followed a long process of conversation and relationship building. ABCD was introduced to the Office of the Environment through Lorelei Hanson, who had known many of these civil servants for nearly a decade through her citizen-at-large position on the Office of the Environment's Environmental Advisory Committee. The Office of the Environment had extensive experience developing and shepherding controversial policies through city bureaucracies, and in building public support and managing public concerns about environmental policies and strategies. They had consulted extensively with many stakeholders in shaping and framing their latest environmental strategic plan,

The Way We Green (TWWG), and indeed, climate-oriented elements of TWWG arose out of a stakeholder consultation. But these civil servants struggled to build civic and institutional support for TWWG, particularly its ambitious climate change and energy objectives.

Divergent interests brought each partner to the table. The Office of the Environment's reasons for partnering were highly pragmatic: they hoped that strong citizen engagement would produce information that would help them align energy and climate policy, already fairly advanced in development, with the perspectives of citizens, and to show this alignment in order to secure political support for an energy transition strategy. While ABCD's mandate emphasized research, learning, and capacity building on deliberation, these were not prominent among the expressed priorities of the Office of the Environment, though it was amenable to ABCD pursuing these goals within the collaboration. ABCD's project director also connected with a councillor who held the environment portfolio for the city, and who was passionate about climate change; he was seeking ways for the city to address this pressing and difficult issue, and at key junctures encouraged the Office of the Environment to pursue a partnership with ABCD.

Perceived risks also distinguished the two partners. The Office of the Environment had a great deal riding on the outcome of the deliberation they designed with ABCD: the deliberation was only one part of a climate and energy plan that was a years-long piece of work, and if the citizen engagement component compromised the plan's development it would be a major setback for the unit and a blow to the civil servants who had invested massive amounts of time in its development. Consequently, there were strong incentives for managers from the Office of the Environment to be closely involved in developing the project in order to maximize the benefit and minimize the risk to the bigger plan. Six members of the Office of the Environment were initially involved in the Edmonton Panel project, though the time burdens of the collaboration were intense and one manager from the city's side eventually took the lead, with others coming in to advise on particular aspects like budget and publicity. For ABCD, this project was the first deliberation it was designing and fully managing; there were many months of uncertainty about whether the project would go forward, and ABCD's leadership and members made tremendous efforts to sustain its progress. Three practitioners and two researchers were intensively involved in the development of the project, and many others advised, developed, conducted research, and participated in a range of workshops that ABCD conducted to support the design of the project.

Because of the heavy investment of both parties in having a successful deliberation, and the different risks and desired outcomes each brought to the relationship, there were many points of struggle and negotiation throughout the collaboration. There was a sustained struggle to align each group's understanding of what citizen deliberation meant and how it differed from consultation approaches more familiar to city managers. ABCD researchers and practitioners did their best to communicate the rationale and principles underlying citizen deliberation, how it could be conducted in a bias-balanced way, and how it could fit productively with the city's decision-making structure; they did this in the course of negotiations about a deliberation process and through workshops organized to educate and engage city staff. Yet prior to the start of the deliberation an Office of the Environment staff member still expressed puzzlement about the approach:

> Some of the tools . . . I'm not comfortable with, but again I'm open . . . I
> know that they will resonate with some people, but I'm just wondering if
> that makes [the citizens] informed or that makes them influenced. (KI 1-4)[1]

Another civil servant remarked late in the partnership:

> Yeah, sometimes you know . . . and even when we thought we were on the
> same page, we weren't on the same page because the language we were using
> . . . you know . . . we'd understand it differently. (KI 2-3)

Some design elements for the Edmonton Panel proposed by ABCD felt inappropriate or too risky to city staff. For example, proposals to have diverse stakeholders address panelists directly or to have a councillor deliberate alongside panelists were felt by one Office of the Environment staff member to compromise the objectivity and representativeness of the exercise (KI 1-10). Tensions also arose from understandings, or misunderstandings, of ABCD as a research-oriented project. Office of the Environment staff at times seemed to perceive deliberative democracy as an obscure, scholarly concept, and ABCD as principally interested in generating research and lacking a sufficient grasp

---

1 Ten members or associates with either ABCD, or the City of Edmonton, who had some role in the Edmonton Panel were interviewed at the start of the citizen deliberation (1) and following it (2). To protect the anonymity of the key informants (KI) they have all been assigned a number ranging from 1 - 10 (e.g., KI 1-5 designates this is an interview with the fifth key informant before the deliberations).

of the difficult political negotiations typically faced by their department. In the words of one civil servant:

> ABCD, I mean we're kind of funding part of their education. I'm sure they're looking forward to publishing and whatever else, but my eye is on the prize that I have to live with after they're gone, [which] is a quality outcome. It seems a little bit like a second priority, like the academic pursuit and brilliance and, you know, academics is number one, and oh yeah, you have this deliverable which sounds very businessy. But we've got to take this to Council after and it's actually got to have value. (KI 1-10)

It took a year and a half of negotiation and planning with ABCD for the Office of the Environment to feel confident enough to move forward with the Edmonton Panel. Key design issues that had to be negotiated included: gathering a demographically representative and attitudinally diverse group of citizens; supporting citizens' learning about climate change and policy choices facing the municipality; facilitating deliberation and voting on recommendations; and producing a citizen report for the city. While city staff trusted ABCD practitioner members' facilitation expertise, they were unwilling to delegate design of the citizen deliberation to these experts; rather, particular decisions were hashed out, and often revisited numerous times. This level of co-design differs from other approaches to deliberation planning, such as the America*Speaks* model, which preserved an arms-length relationship with funders and those to whom citizens made recommendations, securing more autonomy than ABCD had in project design and delivery (Lukensmeyer 2014). ABCD needed to be responsive to the needs of city partners, but this meant that ABCD's team had less clear authority to determine certain aspects of the work. For example, a disagreement arose with the Office of the Environment regarding the presence of media at the deliberation, which would have included a press launch event, press access to the Edmonton Panel, and public sharing of session reports.

While there was initial agreement about some limited publicity for the event, a few months before the deliberation Office of the Environment staff argued against involving media, suggesting, among other things, that it would undermine the objectivity of the citizen deliberation. City staff also worried that media attention in advance of the panel's formal recommendations to City Council would lead some councillors to feel pressured by the administration. This was an unwelcome change of position from ABCD's point of view: based on the experience of practitioners, and knowledge of other deliberative exercises (Parkinson 2006; Cutler et. al. 2008), ABCD saw strong connections between public awareness of citizen processes, the public legitimacy of these exercises, and a sense of accountability on the part of elected officials to attend seriously to recommendations. However, the Office of the Environment was inclined to assign the deliberation process a more limited role in policy or project development than ABCD desired. ABCD members felt they were offering a promising alternative pathway to building public support for the energy and climate change plan, and were frustrated by what they viewed as the Office of the Environment's attachment to more familiar and manageable practices of public engagement.

Notwithstanding the struggle involved in aligning ABCD and Office of the Environment objectives, working together over months built trust and understanding that resulted in a successful citizen deliberation and sustained commitments over several years from members of ABCD and the Edmonton Panel, who acted as champions for the energy and climate change plan that the Office of the Environment was taking to City Council. Members of ABCD, who at times were impatient with the caution demonstrated by their city partners, observed Office of the Environment staff doggedly working to make environmental policy changes in difficult circumstances, and consequently developed empathy for them and admiration for their efforts. Office of the Environment staff saw ABCD members doing effective work during the deliberation, and came to appreciate the voice they were given in decision making about deliberation, as seen in these remarks by two members of the Office of the Environment:

> I was happy with that opportunity to be listened to . . . to be sort of tolerant of the fact that we weren't experts in this field and that we were maybe asking dumb questions and perhaps being a little bit anxious at times when maybe there wasn't a need to be in retrospect, but again I think [ABCD] were patient with us. I think you know they listened to us and they gave us an opportunity to influence the process so I think it worked well. (KI 2-4).

> This was a complex project, and I can't imagine it being done in a better way. I think if it was a cookie cutter, we've done this 50 times, and hired a consulting firm, but that's not what this was about. And so I think it was about learning together and I think we weren't always the easiest client to have and in those times David was tolerant, and at times where we needed to be pushed because we were worried about the timing of things, there were times where he read us the riot act and that was a good thing to do. . . .

It was a great outcome. We invented our way together. . . . And it became a good team effort as we got to know each other. (KI 2-8)

Especially as the deliberation approached and in the eight weeks over which it was carried out, there was a sense of being in it together, and also a willingness to give and take. As one Office of the Environment staff member recollected after the deliberation, "I think what really worked well is a willingness . . . for everybody wanted the panel to succeed, whatever the outcome of the panel was . . . . I mean everybody was passionate about that and willing to work toward that . . . I mean work through the challenges" (KI 2-3). Moreover, as the deliberation progressed, the views of citizens ended up supporting the approach favoured by the Office of the Environment, and the deliberation process as a whole became celebrated by members of ABCD, city staff, and elected officials.

Case Study 3: A Virtual Deliberation on Energy Efficiency Choices

The AEEA, a network of industry and NGOs working to advance energy efficiency in Alberta, was interested in collaborating with ABCD. AEEA's executive director had been involved in ABCD's work since its inception as a representative of the Pembina Institute, a Canadian environmental NGO. When ABCD launched a funding competition in 2012 for deliberation projects led by members of the broad network, AEEA's executive director joined with one of ABCD's deliberation practitioners to propose an online and telephone-based citizen deliberation on Energy Efficiency Choices. The objective was to use the deliberation as a citizen education tool and to support AEEA's lobbying efforts with the Alberta government by demonstrating citizen support for provincial energy efficiency incentives and funding (see chapter 7). Even though climate change was not a dominant frame, members of ABCD were interested in the project, as it would provide a comparative case study for the Edmonton Panel, and also extend ABCD's experience with designing and facilitating different citizen deliberation formats.

While AEEA worked with one of ABCD's deliberation practitioners to design and carry out the deliberation, there was limited collaboration between the AEEA and ABCD teams. There was a research component in the online deliberation proposal, but this was not a major part of AEEA's desired outcomes. ABCD researchers participated in several initial planning calls, and their questions and suggestions were met with ambivalence and concern from AEEA's executive director. Given the limited funding AEEA had secured from ABCD to finance the deliberation, he was concerned that the time investment required to align research with the deliberation would overburden AEEA; as a contractor, the executive director said he could not afford to devote hundreds of unpaid hours to a research project without clear results that would enable both him and his organization to achieve their objectives. As well, he was concerned that research questions being proposed by ABCD might compromise AEEA's use of the deliberation outcomes to demonstrate to government that Albertans supported increased funding for energy efficiency and incentive programs. Perceiving the research as introducing an element of risk, the AEEA executive director gave limited weight to feedback he received from ABCD's researchers on both the design of the deliberation and interpretation of the data.

The development of the deliberation, which took place over about ten months, resulted in a significantly less robust collaboration than that developed with the City of Edmonton. The deliberation design was handled on a largely consultant-client basis between its two leads, and researchers struggled to gather data associated with this fast-moving project. Communication challenges arose when ABCD leadership and researchers sought to understand how AEEA intended to use the outcomes of the deliberation, and whether and how the final report would be made public. While the two leads worked well together, relationships of trust were never developed between AEEA and the ABCD researchers, which impeded collective learning.

In the end, the deliberation had mixed and somewhat confusing results. As "a very instrumentally framed dialogue" with a constrained time frame of only two hours, the deliberation did not allow citizens to "explore the complexities of climate change" as it related to energy (Haas Lyons 2015). The deliberation's final report was not made public or delivered to government because of AEEA's strategic judgments about timing and the usefulness of the results in achieving their political objectives. However, in spite of these challenges, the AEEA executive director found some of "the results were quite useful" in his discussions with government officials and provided "a powerful message" about citizens' interest in energy efficiency (Row 2015). As well, the distributed deliberation allowed for key learning about the use of online technologies to engage citizens in a deliberation.

Case Study 4: A One-Day Deliberation on Water in a Changing Climate

ABCD's 2012 competition funded a second deliberation proposal from an ABCD researcher/practitioner team to conduct a deliberation in conjunction with the OWC. The deliberation practitioner had already done consulting work with the OWC, and the researcher had met the executive director of the OWC after delivering a public presentation. OWC had considerable experience in undertaking citizen engagement but deliberation was new to them. The OWC did not have a policy moment that the deliberation could be linked to, nor did their work directly address climate change; but they had an interest in exploring how they might use citizen deliberation in their education and stewardship activities related to land and watershed management to engage citizens on climate change (see chapter 5). ABCD was willing to fund the project because it provided a comparative case study that could extend ABCD's social learning about citizen deliberation on climate change. The project was called Water in a Changing Climate.

The design and development of the deliberation project were led by the ABCD researcher, who hired a project manager to assist her and consulted with the deliberation practitioner around design. The objective of the deliberation was three-fold: to engage with communities outside of Alberta's major metropolitan centres; to focus on a different aspect of climate change than energy; and to see what kind of deliberation could be accomplished in a day (Blue 2015). OWC was not heavily invested in the outcome of the deliberation; its leaders were not active in the design of the event and assumed a largely advisory role in the project's development. This gave the ABCD members scope to design the day-long event in ways that met their interests, particularly the researcher's interest in framing climate change in terms of adaptation rather than mitigation.

As was the case with the AEEA deliberation, there was a lack of communication between the ABCD members overseeing the Water in a Changing Climate project and the rest of the ABCD team. This breakdown in communication resulted in a lack of data being collected and confusions regarding how the deliberation outcomes would be used by the OWC. Members of ABCD not directly involved in overseeing the deliberation had hoped the OWC would use the results to inform its work, and did not grasp until long after the event that OWC, in understanding the deliberation as a pilot exercise, therefore felt no need to engage directly with citizens' recommendations (Frank 2015). The executive director of the OWC saw the value of citizen deliberation as "a way to explore issues and solutions and be open-minded," rather than a way of shaping organizational direction in a more determinate way (Frank 2015). These tensions and gaps around data gathering and research outputs occurred, as with the Food and Agriculture deliberation, primarily between academics, and reflected not only challenges of busy-ness and communication over distance but a failure to settle collaboratively, clearly, and early on the priority of different research outcomes.

There was significant learning about deliberation by particular individuals, but as was the case with the Energy Efficiency Choices deliberation and the Food and Urban Agriculture Citizen Panel, there was a lack of collective learning across ABCD. ABCD's leadership struggled to communicate with the project team about the development of the project and its results, and to find ways for other ABCD researchers to collaborate on the project. Nevertheless, the Water in a Changing Climate deliberation generated several worthwhile learning opportunities. In addition to learning by citizens, the OWC found the results of the project educational in introducing to them to the deliberation process and the range of citizens' views on climate change (Frank 2015); the lead researcher spoke of "huge learnings" about project management, deliberation, framing, facilitator training, exploring values, and discussing climate change (Blue 2015); and the practitioner expanded her knowledge of the challenges associated with engaging citizens for a one-day deliberation on climate change (Dale 2015). However, for shared learning to have occurred, these individual experiences and reflections would have needed to go "beyond individuals or small groups to become situated within wider social units or communities of practice" (Reed et al. 2010), which proved a challenge for this project.

A Comparative Analysis

Looking across the four cases, we see several factors that, when combined, support successful collaboration in deliberation projects.

Not unexpectedly, the primary factor that led to deep collaborative relationships was the development of trust. Trust revealed itself as respect for different contexts and cultures of risk in relation to the collaboration and the deliberation being planned, and it enabled organizations with different reasons for partnering to open up their agendas to align their activities and respective trajectories of work in ways that met not only their needs but also the other organization's needs. Establishing trust helped to build understanding, as well as to support tolerance for misunderstandings that resulted from different institutional and professional cultures, incentive structures, roles, contexts, and paradigms. As the project manager from the City of Edmonton noted:

> There is a risk in not knowing these people, not knowing the process, not knowing where this could go, how competent they are, how capable they are, whether we are wasting our time. So those are the sorts of risks we were thinking about. But in terms of our relationship with the different individuals, it evolved over time, and as we became comfortable with their abilities, their competence, their methodologies, the team they were using, their commitment, their professionalism, the comfort level grew. Yeah, and the relationship grew as well. (Andrais 2015)

Mutual trust allowed the negotiation of differences of opinion on how citizen deliberation could play a role in decisions, and even on who "citizens" were in relation to policy and program development; these interpretations are important as they define and constrain the development, form, and outcomes of public dialogue practices (Newman et al. 2004). It also enabled the negotiation of different goals and strategic objectives in relation to the partnership, and different orientations to research and learning.

This is not to say that divergences were always expressed or resolved. While certain members of ABCD sensed uneasiness among some of our partners over the course of the development of deliberations, this was difficult to address directly as such matters were often felt to be internal and political, not ones for public discussion. Trusting relationships enabled collaboration to proceed in spite of these tensions; but this is not to say that trusting relationships bring to light or settle every difference, or that they have to.

Several secondary factors also shaped the quality of the collaborative relationships. First, the most successful collaborations arose when ABCD's outside partners had a strong investment in outcomes of deliberation: they had something at risk, and a motivation to work collaboratively to manage risks and produce the best outcomes. At times, because of this major investment in outcomes there was a strong incentive for the outside partner—with the capacity to influence decisions—to jump to its own solutions to problems, so there were seeds for disagreement, but also the capacity and incentive to process these disagreements communicatively and collectively. Working through misunderstandings and disagreements, when done effectively, can build or reinforce respectful relationships and mutual trust, which in turn can increase commitment to the deliberative process and to social learning. It also can clarify terminologies, commitments, and goals in ways that lead to individual and social learning, and more successful deliberative processes and outcomes.

A comparison between Edmonton's Energy and Climate Challenges deliberation and the Water in a Changing Climate project demonstrates how strong interests in the deliberation outcomes translated into deeper collaborations and more useful citizen recommendations. Both the Office of the Environment and ABCD had strong investments in the Edmonton Panel. The project attracted the most commitment from researchers and practitioners in ABCD, and there was involvement from many civil servants. While there were many ups and downs in the development of the project, disagreements and misunderstandings were hashed out in meeting after meeting, building trust and mutual understanding over time, as illustrated by this interview excerpt:

> David [and the lead facilitators] were new to us. And that relationship changed and evolved. We watched them and watched how they delivered on their vision, on their program, and so everything we saw was positive and it just convinced us more and more that these were people who were competent, who were committed, who had passion in what they were doing and what they were delivering. In my experience over decades not everybody delivers . . . sometimes they don't deliver everything they promised, and sometimes expectations are less than what you'd hoped, and in this case the expectations exceeded, or the performance exceeded our expectations right across the board. (Andrais 2015)

In part because of the strength of relationships and trust, project development was given the time it needed to coalesce around a mutually acceptable deliberative exercise. The influence of the Edmonton Panel on city decision making and the uptake it received from elected officials in public debates themselves supported the breadth of learning from the process that is necessary for social learning to occur.

In contrast, the owc had limited investment in and risk associated with the outcomes of the day-long Water in a Changing Climate event, as they were involved out of a general interest in learning about deliberation rather than an intention to use the outcomes politically. While the owc did play a key support role in the citizen deliberation, including the provision of small group facilitators and note takers, they did not engage significantly in the planning and design of the dialogue process because they were not looking for any specific kind of feedback or responses to a particular issue. As a result, some of the recommendations that arose out of Water in a Changing Climate were very general and "outside of the purview of the Oldman Watershed" (Blue 2015).

Second, collaborative relationships were strengthened when each organization's goals were well understood by both parties as a basis for defining shared goals. One effect of this joint contribution and ownership was that information could be shared more broadly, rather than small numbers of players having the prerogative to restrict the circulation of information or research data; trust, communication, and effective collaboration diminishes when such information can be restricted. In hindsight, we see how, in our collaborations, shared goals needed to constantly be revisited and revised. For example, AEEA was invested in the outcomes of the deliberation designed with ABCD, but ABCD's commitment to deliberation projects that were strongly connected to decision making and to sharing outcomes regardless of how they align with the policy views of a partner created unease for AEEA's executive director. As a result, given that ABCD researchers were in an ancillary role, decisions were made by AEEA, without consultation with researchers, which adversely affected the collection of quality research data. Time pressures were a factor here too, but more significant was a lack of incentive and structures for collaboratively working through challenges, including around the fit between research and practice.

Notwithstanding the different learning outcomes of the three deliberations that ABCD helped to run, there was in each instance a shared commitment to both strong environmental responses and good citizen involvement; this shared commitment provided an explicit and common basis for each partnership and associated deliberative processes. Office of the Environment staff, the AEEA, OWC, and ABCD members shared a commitment to effective climate action, and there was a sense of solidarity around this. Even though we diverged at times on how good work with citizens would contribute to effective climate policy and action, our shared values around these objectives saw us through tensions and conflicts. This held true within ABCD as well.

Third, trust and shared learning critical to collaboration were strengthened when time and energy were devoted to appropriate and respectful communication about risks and goals (Beattie and Annis 2008; Suarez-Balcazar, Harper, and Lewis 2005). The hurried and pressured circumstances of collaboration often led us to neglect matters like checking in to clarify confusing situations or ask about views and preferences, or carefully exploring what we each needed to get out of the collaboration to feel satisfied and properly supported. Paying more attention to these issues might have supported greater reciprocity and relationship building, which would have lent resilience to our best collaborations; it might also have helped ABCD researchers to notice when we were relating to others in the room as obstacles or objects of persuasion rather than partners in a jointly developed venture. For example, the Water in a Changing Climate project was developed by only one researcher and practitioner within ABCD, with a couple of OWC contacts advising. Researchers outside of this small team were not significantly involved in planning, and when communication broke down around the planning and research, there were limitations to the relationships and collaborative mechanisms that might have resolved these in a mutually acceptable way.

Fourth, the availability of time influenced the success of collaboration. The development of Edmonton's Energy and Climate Challenges deliberation took eighteen months, whereas the other three cases were each completed from start to finish in less than a year. Collective problem solving can be more difficult, and improbable, in projects where there is intense time pressure, which produces the temptation to delegate work, take shortcuts around collaboration, and limit communication and sharing of project details. The research collaboration with CPI demonstrated well how a lack of time easily undermines a potentially rich collaboration. Even had CPI been interested in partnering more extensively with ABCD, the time pressure they were under to develop research tools and design and execute the deliberation left them with little time or inclination to allow ABCD to shape the common project. Consequently, when disagreements arose, given the limited trust and relationships that had been built, the collaboration largely fell apart. While some useful individual learning took place, it was truncated by unwillingness to share authority in crafting research tools, or to share data after the fact.

As we look over the four cases, we also notice that prior relationships had *less* salience than one might have supposed. All four of our cases involved prior relationships between ABCD members and key figures in each outside partner, but these relationships do not seem to have influenced outcomes as much as the partner's degree of investment in outcomes, and the relationships and levels of trust developed or sustained by collaborative decision making over time.

Conclusion

The literature on collaboration shows the importance of diverse stakeholders, common goals, and objectives, well-designed group processes, learning that goes beyond individuals to a community of practice, shared responsibility and power, and uptake by key officials. In this chapter we have used four ABCD cases to foreground important variables in the development of collaborations around citizen deliberations that support the development of trust capable of bridging different organizational needs and objectives; strong investment by outside partners in outcomes of deliberation that can motivate ongoing negotiations of differences; open communication that flows from both shared goals and an understanding of divergences; and sufficient time to address misunderstandings and resolve disputes collaboratively.

ABCD worked with partners to convene citizen deliberations relating to climate change, a quintessentially wicked problem, and to build shared learning out of these collaborations. While many public participation exercises are advertised as achieving such outcomes, our partners at the City of Edmonton reminded us that this is not always the case:

> You see all sorts of engagement efforts and you think a lot of them end up in that same sort of bag of checking a box, a conversation, sort of superficial, never really getting into a deep dive to understand the trade-offs associated with whatever the issue is. What you are doing when you are bringing people together to talk about a tough issue is that you are talking about change, and that change has a range of implications and a range of trade-offs. And so that is what we were able to do in this exercise. (Andrais 2015)

We hope that our reflections on the successes and challenges of these collaborations offer inspiration and also encourage appropriate vigilance in researchers or practitioners planning other deliberation projects to address our toughest problems.

References

Andrais, Jim. 2015. Telephone Interview with Lorelei Hanson, December 9.

Beattie, Marian, and Robert C. Annis. 2008. "The Community Collaboration Story: Final Report." Accessed October 23, 2015. https://www.brandonu.ca/rdi/publication/community-collaboration-story-2008-a-ccp-model-project/.

Blue, Gwendolyn. 2015. Telephone Interview with Lorelei Hanson, June 8.

Cavanagh, Fiona. 2015. Telephone interview with Kristjana Loptson, June 5.

Cestero, Barb. 1999. "Beyond the Hundredth Meeting: A Field Guide to Collaborative Conservation on the West's Public Lands." *Sonoran Institute*, July. http://sonoraninstitute.org/component/docman/

doc_details/1193-beyond-the-hundredth-meeting-a-field-guide-to-collaborative-conservation-on-the-wests-public-lands-07011999.html?Itemid=3.

Chrislip, David D. 2002. *The Collaborative Leadership Fieldbook: A Guide for Citizens and Civic Leaders*. San Francisco: Jossey-Bass.

Cutler, Fred, Richard Johnston, R. Kenneth Carty, André Blais, and Patrick Fournier. 2008. "Deliberation, Information and Trust: The British Columbia Citizens' Assembly as Agenda Setter." In *Designing Democratic Renewal: The British Columbia Citizens' Assembly*, edited by Mark E. Warren and Hilary Pearse, 166–91. Cambridge: Cambridge University Press.

Dale, Jacquie. 2015. Telephone Interview with Lorelei Hanson, May 10.

Frank, Shannon. 2015. Telephone Interview with Lorelei Hanson, July 19.

Haas Lyons, Susanna. 2015. Telephone Interview with Lorelei Hanson, May 14.

Keen, Meg, and Sango Mahanty. 2006. "Learning in Sustainable Natural Resource Management: Challenges and Opportunities in the Pacific." *Society and Natural Resources* 19: 497–513.

Lampe, David, and Marshall Kaplan. 1999. "Resolving Land Use Conflicts Through Mediation: Challenges and Opportunities" Cambridge, MA: Lincoln Institute of Land Policy. http://www.lincolninst.edu/pubs/59_Resolving-Land-Use-Conflicts-Through-Mediation.

Lukensmeyer, Carolyn. 2014. *Bringing Citizen Voices to the Table: A Guide for Public Managers*. San Francisco: Jossey-Bass.

Mattessich, Paul W., and Barbara R. Monsey. 1992. *Collaboration: What Makes It Work. A Review of Research Literature on Factors Influencing Successful Collaboration*. St. Paul, MN: Amherts H. Wilder Foundation.

Newman, Janet, Marian Barnes, Helen Sullivan, and Andrew Knops. 2004. "Public Participation and Collaborative Governance." *Journal of Social Policy* 33(2): 203–23.

Parkinson, John. 2006. *Deliberating in the Real World: Problems of Legitimacy in Deliberative Democracy*. Oxford: Oxford University Press.

Reed, Mark S., Anna C. Evely, Georgina Cundill, Ioan Fazey, Jayne Glass, Adele Laing, Jens Newig, Brad Parrish, Christina Prell, Chris Raymond, and Lindsay C. Stringer. 2010. "What Is Social Learning?" *Ecology and Society* 15(4): r1. http://www.ecologyandsociety.org/vol15/iss4/resp1/.

Row, Jesse, 2015. Telephone Interview with Lorelei Hanson, Deborah Schrader, Mary Pat McKinnon, and David Kahane, June 7.

Schusler, Tania, M. Daniel J. Decker and Max J. Pfeffer. 2003. "Social Learning for Collaborative Natural Resource Management." *Society and Natural Resources: An International Journal*, 16 (4): 309–326.

Suarez-Balcazar, Yolanda, Gary W. Harper, and Rhonda Lewis. 2005. "An Interactive and Contextual Model of Community-University Collaborations for Research and Action." *Health Education and Behavior* 32(1): 84–101.

Weber, Edward P. 2000. "A New Vanguard for the Environment: Grass-Roots Ecosystem Management as a New Environmental Movement." *Society and Natural Resources* 13: 237–59.

Williams, Ellen M., and Paul V. Ellefson. 1996. "Natural Resource Partnerships: Factors Leading to Cooperative Success in the Management of Landscape Level Ecosystems Involving Mixed Ownership." Staff Paper Series 113. Department of Forest Resources, University of Minnesota. https://www.forestry.umn.edu/sites/forestry.umn.edu/files/Staffpaper113.pdf.

7

On the Ground

Practitioners Reflect on ABCD's Citizen Deliberations

Mary Pat MacKinnon, Jacquie Dale, and Susanna Haas Lyons

This chapter reflects the vantage points of three citizen deliberation practitioners deeply involved in the design and facilitation of three Alberta Climate Dialogue projects: the Citizens' Panel on Edmonton's Energy and Climate Challenges (Edmonton Panel); Water in a Changing Climate (wcc); and Alberta Energy Efficiency Choices (aeec). Jacquie Dale and Mary Pat MacKinnon were co-designers and co-facilitators of the Edmonton Panel, Jacquie Dale designed and facilitated the wcc, and Susanna Haas Lyons did the same for aeec. The first two forums were in-person deliberations, while the third was conducted via a web conferencing system, with optional telephone access, and enabled participants to meet in both small groups and plenary sessions. As practitioners, we collectively bring over fifty years of experience in designing, facilitating, analyzing, reporting on, researching, and evaluating deliberative dialogue on many complex topics, for a range of different purposes (including policy, community development and action, and citizen learning), and using a variety of formats and technologies.

Deliberation theorists claim that citizens have a right and a responsibility to be active participants in democracy (Gutmann and Thompson 2004), are capable of doing so (Warren and Pearse 2007; Rose 2007; Lukensmeyer 2012), and that the act of participation builds their capacity (Woodruff 2005; Nabatchi and Leighninger 2015; Prikken, Burall, and Kattirtzi 2011). Climate change discourses and practices have also enshrined the role of citizen participation

(UN 1992, 2015). As practitioners who advocate for deliberative democracy, our design and facilitation approaches to all three deliberations were grounded in these normative (i.e., in the sense that citizens have a legitimate right to participate) and theoretical (i.e., concerning citizen capacity and skills building and the role of values in deliberation) constructs.

Our approach to facilitation of the ABCD deliberations was also guided by what Carl Rogers calls the three core conditions for facilitative practice: realness, acceptance, and empathy (Smith 1997, 2004). We strive to attend to participants' emotional and learning needs. As facilitators, we have a responsibility to assess and balance the quantity and intensity of learning and deliberation requirements with citizens' needs for a safe, constructive, and interesting environment to do their work.

In this chapter, we examine three dimensions of central importance to deliberative practice: issue framing, planning for mini-public deliberations, and enabling citizen deliberation.

> *Issue framing in climate change deliberations*: Framing is about determining how an issue or problem is presented and structured. How to frame an issue is a key decision in planning a climate change (or any) deliberation (Barisione 2012; Kettering Foundation 2011). Issue framing influences what and who are included in a deliberative process, how participants are invited to engage with the issue, the scope of the dialogue, and the range of actions being considered (see chapters 2 and 5). Of course, who or what is leading the issue framing is also critical. In this section, we explore the challenges of framing a hugely complex issue like climate change, and the importance of context, opportunity for policy impact, and our partners' expectations and optics about the framing process.

> *Planning for mini-public deliberative dialogues*: Mini-public is a term used for deliberations that bring together, either virtually and/or face-to-face, a limited number of people who reflect certain characteristics of the broader population. In this section, we explore the particular challenges of recruiting and preparing participants for such deliberations as well as the impact that the partners' goals had on the methodologies employed in the three ABCD projects.

Enabling citizen deliberation: Designing and implementing deliberation for and with citizens brings to the fore certain considerations. This section explores four of these: the importance of values as a critical contribution of citizens; the need for and role of topical knowledge; respect for participant diversity, including perspectives, ways of knowing, experiences, values, education, and ideological world views; and the challenges of citizen ownership and group power dynamics.

Throughout the chapter, we share the challenges that emerged during these deliberations, how we responded, and what we see as strengths, gaps, and remaining questions. The conclusion highlights some key learnings and issues meriting further consideration by practitioners, academics, and decision makers.

Issue Framing in Climate Change Deliberations

Challenge: Complexities of Climate Change Framing

In our role as designers and facilitators, we were very aware that climate change, as a topic for deliberation, presented layered complexities. As a super wicked problem (Levin et al. 2012), efforts to simplify climate change are confounded by its uncertain, unpredictable, indeterminate, and interdependent nature. Furthermore, climate change operates on a time scale beyond the experience of decision makers and citizens alike, making it difficult for citizens to grapple with for many reasons: it is distant from everyday concerns, it is hard to disentangle causation and consequences, and it requires holistic actions involving multiple actors across jurisdictions.

How do we work to help citizens to deliberate on climate change-related issues without becoming so immobilized that they feel powerless to effect real change? We looked to climate psychology for some guidance on framing the issue. There is a substantial and growing body of social science research suggesting that it is counterproductive to present climate change primarily in terms of fear and dire threats to the globe, which can create participant paralysis or anxiety. A more positive orientation is recommended where citizens see themselves as agents of change who are able to overcome hopelessness or fatalism (Pike, Doppelt and Herr, 2010; Goleman 2013, ch.14).

In all three ABCD projects, we explored climate change from more local or provincial perspectives: the Edmonton Panel initiative was at a city scale, the

WCC deliberation focused on a watershed view, and the AEEC deliberation had a provincial focus. The strength of this approach was that people explored what climate change meant for them and what they could do about it in their local contexts.

Climate change is an emergent issue; despite the best modelling, we cannot know with certainty what it will look like in communities in twelve, twenty, or fifty years. This uncertainty requires a participatory approach that supports people in engaging over the long term in taking responsible action in their communities. We focused on ways of empowering participants to deliberate on what communities can and need to do to address the intractable issues of energy transition and climate change, taking an iterative approach to this task. The in-person deliberations included materials and presenters on the national and global dimensions of climate change, but given our framing and time constraints this input was limited. The online deliberation referred to the relationship between energy use and climate change, but it did not elaborate on the complexities of climate change. As well, we had to invest time to become more familiar with the critical research and arguments relevant to the Alberta context.

Making the issue of climate change manageable and something we could tackle with limited time and resources also meant examining a defined set of attributes such as energy resiliency, and ignoring other attributes and responses, such as the impact of climate change on species. These framing approaches were developed collaboratively with our ABCD convening partners and revolved around issues that were core to their mandates and/or to a specific policy opportunity. As a result, it sometimes felt as if we were focusing on things not normally connected to climate change, or of minor impact given the global scale of climate change, for example, energy efficiency for a city vehicle fleet. However, in order to advance, we need to limit our examination of climate change to the lenses of energy transition, energy efficiency, or water management.

Challenge: Policy Context and Convening Partners' Perspectives

Deliberation context encompasses the policy context, setting, and relationships in which deliberation occurs, and the powerful reality that these always matter in designing and delivering citizen deliberations (Abelson and Gauvin 2006). This is even more the case when the deliberation topic involves climate change. As chapter 3 of this volume describes, all three deliberations took place in Alberta, a province whose wealth has been historically heavily dependent on fossil fuels.

When opportunities arose to host a deliberation that could potentially influence a live policy decision, ABCD was keen. This was the impetus for the longest and most expensive deliberative undertaking—the Edmonton Panel—as well as the AEEC project. But these policy opportunities came with an inherent framing for discussion and trade-offs about what to exclude, or at least to minimize, in the deliberation.

The opportunity presented to ABCD offered by the City of Edmonton to shape policy heavily influenced our willingness to trade off a narrower scoping of the Edmonton Panel deliberations than might have otherwise been the case. The policy options were contained in the city's Energy Transition Discussion Paper, a background document written by energy efficiency experts that details three scenarios for energy transition within Edmonton. Our city partners were looking to citizens to provide guidance on which path to follow and which policy levers to pull. The discussion paper was technical and assumed a fairly high level of literacy and understanding of policy and science. As the lead guide for citizens, the discussion paper shaped the framing and the content for deliberation in significant ways. For example, the implications of Alberta being an energy powerhouse dominated by its carbon-intensive oil sands were not a focus of discussion. Also, Alberta's large carbon footprint, predominantly a result of the extraction, processing, and transportation of fossil fuels, was not a significant aspect of the citizens' deliberations apart from the discussion around the provincial energy grid's heavy reliance on coal.

Energy use also came to the fore in the AEEC deliberations. ABCD's partner in this exercise, the Alberta Energy Efficiency Alliance (AEEA), felt that a discussion of energy efficiency, as opposed to other aspects of climate change, would have the greatest potential to influence provincial policy. The AEEA identified two areas where informed public input would be most useful for its engagement with government actors: energy efficiency regulation and funding. The first portion of the deliberation asked participants to consider options for funding provincial energy efficiency programs. The second portion considered if and how new energy efficiency regulations should be established. Some participants appreciated the opportunity to provide input on issues of importance to government, while other participants struggled with and objected to the narrow framing. As one AEEA participant remarked, "the focus was so narrow but it allowed for an in-depth consideration of one part of a bigger picture."

A different approach was taken in the WCC deliberation, as there was no immediate policy opportunity around which to focus citizen input. ABCD's partner in this event was the Oldman Watershed Council (OWC), a not-for-profit multi-stakeholder group. Since the OWC's interest was more in the process than the outcomes of the deliberation, the framing for this one-day deliberation was largely in the hands of the ABCD team. Water was chosen as the broad issue for deliberation, which aligned well with OWC's focus on its watershed. Within this broad framing, citizens identified their concerns about climate change and water in their region and clustered these into themes, which then became the topics for deliberation.

How We Responded

In the Edmonton Panel, the discussion paper's technical and more circum-scribed framing was counterbalanced in several ways: a participant handbook was developed to provide additional information on and framing of climate change, and a variety of presenters and resource people introduced a range of perspectives on climate change. Early in the process, we took citizens through a discovery of their own values in relation to energy transition and climate change, asking them to identify what values they believed should guide deliberation on the recommendations; and we continually invited participants to consider their own and others' interests and perspectives.

On balance, the city's objectives did not stymie citizen deliberation. Citizen panelists were able to place their deliberation in a broader context of climate change while still feeding directly into a policy opportunity, which for many made the deliberation more meaningful. For example, the participants urged the city to "Go faster, Go further," emphasizing that the city must set strong, measurable targets for energy transition in a five-year time frame (CPEECC 2013). And, even though the technical overlay and limited scope of the dis-cussion paper constrained the panelists and resulted in some frustration, it also led to creative thinking as participants found ways to introduce new ideas or priorities, such as fiscal prudence and sensitivity to the vulnerabil-ity of lower-income Edmontonians to cost increases. Allocating time for and encouraging emergent thinking were essential design approaches to enable these kinds of outcomes.

The AEEC deliberation focused on issues the Government of Alberta was considering action on but on which it wanted guidance about value-informed choices, such as who should shoulder the financial burden of energy efficiency programs and what sectors should be regulated. Even with this constrained agenda, small group discussions were designed so that participants could raise related issues they cared about and pose questions (typed or spoken) during plenaries. As well, the training session for small group facilitators encouraged them to be responsive to participants' interest areas. In practice, however, challenges associated with the technology used to allow participants to join remotely, such as audio quality, and the constrained timelines meant that some discussion groups had difficulty getting beyond the assigned task to more deeply explore trade-offs or related issues. Plenary sessions were more responsive to participant agenda setting through convenor/participant exchanges in the online chat area, and oral question and answer exchanges.

The WCC deliberation was intentionally designed as a counterpoint to the Edmonton Panel deliberation in some key ways. ABCD was interested to see what could be accomplished in a one-day session with a more limited budget. The interest, in particular, was to experiment with an approach to deliberation that not-for-profits and communities could easily take on. The day-long session included a presentation on climate change in southern Alberta, the identification and clustering of concerns into issue areas, and subsequent deliberation on these areas of focus.

Strengths, Gaps, and Questions

The policy opportunities and the community contexts within which the projects were conceived, combined with the limited time and resources available, shaped the deliberations and posed challenges and opportunities for us as designers and facilitators. In the Edmonton Panel, we struggled with making the process rigorous enough to be taken seriously as useful policy input by City Administration and Council, while working to make space for emergent and other perspectives from participants. Participants achieved a high level of consensus on the recommendations presented. However, with the notable exception of panelists defining a new recommendation for the city to "Go faster, Go further" on energy transition and carbon reduction in the second last session, they largely confined themselves to the policy options presented in the discussion paper. While the panel members ably performed their citizenship tasks, and the results of the deliberation did have an impact on the subsequent policy decision, we were left to wonder what sparks of creativity that might have resulted in additional or different directions were lost.

In the AEEC deliberations, some participants wanted to discuss energy production in Alberta but were constrained by the focus on energy efficiency regulation and funding and the related opportunity to inform government policy. The narrow framing enabled AEEA to draw on the deliberation results in its engagement with the Alberta government on this topic. Undoubtedly, more time would have provided some flexibility in the agenda for participants to discuss additional issues—such as climate change's global aspects and/or Alberta's energy production. If participants had been invited to discuss these issues, they might have expressed a broader range of ideas and become more interested in the related issues.

The WCC dialogue was rich and rewarding for participants, and the OWC learned much about the process of deliberation, but the ideas developed through the deliberation have not been acted on. According to the OWC's executive director, this is because the Council has not yet made climate change a part of its strategy, and before it would act on the WCC recommendations, it would conduct more dialogues with other people in other communities (Frank 2015). In many ways, the most visible outcome of the WCC deliberation was OWC's and the participants' learning in terms of discovering a new way to talk. When participants framed their concerns into the issues for deliberation, it allowed for diverse and creative topics on the issue of water to surface. This framing was also influenced by the context participants brought into the room. For example, one of the issues participants identified for deliberation was extreme weather events, a top-of-mind concern given the catastrophic flood the region had experienced the summer before the deliberation.

Planning for Mini-Public Deliberative Dialogues

Challenge: Recruit Diverse Mini-Publics

All three of the deliberations ABCD led were mini-publics. We took this approach to move beyond the "usual suspects" (e.g., those who are most vocal and organized and most likely to show up), in order to gather a cross-section of perspectives and experiences that are reflective of the community. Citizens bring different points of view, ways of knowing, experiences, values, education, and ideologies to an issue, and these all influence a deliberation. But mini-publics are challenging from a design perspective. They demand methods and techniques to ensure that all voices are heard, that learning approaches appeal to diverse participants with different learning style preferences, and that power differentials among participants are minimized. Methods and approaches must be designed to bridge participants' different ways of knowing and learning, and to support them in exploring common ground, while protecting space for divergence and differences.

How We Responded

The type of recruitment or selection process used varied with the nature of the opportunity (see chapter 4). For the Edmonton Panel, where our city partners and their senior managers felt it was critical that the panel reflect the diversity of Edmontonians, a more rigorous, statistically valid process was needed than in the WCC deliberation. The City of Edmonton and AEEA wanted the citizen panels to reflect the larger population on a number of demographic measures and mirror diverse attitudes on energy and climate change, and so a public opinion research firm was contracted to recruit participants. However, a professional third party was not used to recruit participants for the WCC deliberation. Instead, recruitment was largely done through public service announcements in local papers, word-of-mouth, and contact lists coming from the OWC. The WCC also employed targeted recruitment, such as for First Nations people, which was largely successful and added to the diversity of perspectives and patterns of discourse.

Strengths, Gaps, and Questions

Lower-budget, shorter deliberations like WCC are more vulnerable to the effects of whatever burning issues or preoccupations people bring into the room with them than longer deliberations like the Edmonton Panel. For example, while WCC recruitment included some attitudinal dimensions, it was not possible to compare the range of participants' attitudes to a broader population poll as was done in the Edmonton Panel deliberation. In addition, while attempts were made to recruit broadly for the WCC, several participants learned about the deliberation from the OWC electronic mailing list and hence were familiar with the Council and its work. This influenced participants' ability in two interesting ways. Some self-selected small groups used their pre-existing knowledge to leapfrog the discussion into new areas, whereas another discussion group had difficulty moving beyond the usual conversation around their issue area.

The AEEC participants were also less diverse than the Edmonton Panel due to drop-off between initial recruitment and sign-up for the deliberation sessions, as well as technological challenges citizens experienced in accessing the deliberation. Only about one-third of those recruited participated, which limited the diversity of views heard. On average the participants had higher levels of education than the typical Albertan. We are left wondering what messages and interactions by both the recruiter and ABCD would have increased turnout. Nonetheless, some participants valued the opportunity to hear others' perspectives, as noted by this AEEC participant: the deliberation "made me aware of others' viewpoints and that they are often very different from mine. I changed my mind on some topics as a result of this input."

Challenge: Align Deliberation Methodologies with Partners' Goals and Resources

Partnerships enriched the design, planning, and facilitation of ABCD's deliberations at all stages, but they also made for more complex processes. The convening teams included deliberation practitioners, government officials, not-for-profit representatives, policy advocates, and researchers. Implementing the deliberations brought together citizen participants, lead facilitators, small group facilitators and note takers, researchers, and administrative support. ABCD's collaborating partners came from different institutional contexts and were driven by different considerations. This impacted and influenced the choices of design and methods used to engage participants, including participant tasks, length of time for and approaches to learning and deliberating, decision-making methods, approach to note taking and small group facilitation, reporting, and more. This presented advantages and challenges for us as designers and facilitators.

How We Responded

The duration of the Edmonton Panel, spanning six Saturdays (early October to early December), was dictated in large degree by available resources and our city partners' view of what they perceived to be the outside limit of citizens' willingness and capacity to volunteer their time. As it turned out, with a few exceptions, the panelists' attendance record was excellent, as was their level of engagement. Another important influence on the structure of the Edmonton Panel was the City's request for detailed session-by-session design and material descriptions in advance of the panel launch. ABCD willingly provided these as a way to instill confidence in the robustness of the process on the part of our city partners and small group facilitators. In reality, each session required new design thinking, additional work, and the creation of new materials, all of which translated into a requirement for just-in-time responsiveness from the on-site team. As lead facilitators and process designers, going into the dialogue we knew from experience that the process needed to be iterative, building on what had happened in previous sessions. We also knew what citizens required for informed, meaningful dialogue, and what interested and energized them (MacKinnon, Dale, and Schrader 2014). We employed design strategies that allowed aspects of the process to emerge, such as using electronic keypad voting for deciding which recommendations to include in the report; using open space methods that allow citizens to self-organize around topics of interest (Owen 1997) in the deliberation; incorporating a climate change psychology presentation in a session; and nimbly shifting agendas and tasks to accommodate participants' diverse energy levels and psychological states.

As the WCC deliberation did not feed into a concrete, specific opportunity, the deliberation was designed to be highly responsive to participants' interests and the areas for deliberation and accompanying values. This approach was beneficial for exploring climate change–related topics through unusual lenses, which has the potential to uncover creative solutions. As examples, participants in the group discussing social justice and responsibility recommended fostering individual stewardship for development of the common good and the group deliberating on environment and human health developed the idea of advocating for and supporting sustainable food production in urban centres.

The AEEC's distributed format enabled people from across Alberta to participate in the deliberation, regardless of location, which helped to meet AEEA's objective of relevancy for the provincial government. Additionally, the cost-effectiveness of the online/phone format was influential in choosing this method over options such as a handful of small meetings in various locations throughout the province. Given these parameters, the AEEC deliberation was designed in response to the idea that citizens wanted to give their input but were likely not willing to participate online for extended periods of time. This translated into involving each participant in a single two-hour online session.

Strengths, Gaps, and Questions

The structures of the Edmonton Panel, WCC, and AEEC each constrained participants' involvement in specific ways. In the Edmonton Panel, we wonder whether having an additional Saturday or two might have resulted in deeper deliberations on critical issues, greater clarity on some of the trade-offs required, and broader articulation of additional recommendations and options. While several of the recommendations resulting from the WCC were novel (Frank 2015), they were not necessarily within the mandate of the OWC, nor was the process designed to generate citizen action. In addition, the emergent nature of the process (in which the areas of focus for the afternoon's deliberation arose from the morning's input), put significant pressures on the small group volunteer facilitators, who did not know what themes they would be working on in the afternoon. With hindsight, this design probably necessitated a higher level of facilitation skill and deliberation experience than some of the facilitators possessed, even with some training the day before the event.

The AEEC's online/phone format enabled a province-wide conversation; however, the short time frame constrained the amount of time participants engaged with the issues and one another. Further testing of the assumptions that participants would be reluctant to join a session that was longer than two hours or attend multiple sessions would have provided some useful information to guide the deliberation structure. In addition, offering an optional and short, advance preparatory session for people new to the technology would have alleviated some of the technological challenges of using an unfamiliar-to-some tool. Modifying the format of interaction is another option that could be considered, such as offering a traditional webinar that primarily presents information to attendees, which could then be followed up with small group discussions via conference call, using a service such as MaestroConference.

Challenge: Support Participants' Learning and Deliberations

Achieving informed participation is widely accepted as essential for good deliberation, but there is no single answer as to what constitutes informed participation and it is not unusual for different partners to have different interpretations of this. For example, in the Edmonton Panel there was a healthy tension between ABCD and the city partners, who were interested in ensuring that participants were as knowledgeable as possible about the science and technological aspects of Edmonton's energy and climate challenges so that their recommendations would be as informed as possible (see chapter 6). While agreeing with the need for informed participation, we were skeptical of the assumption that greater quantities of knowledge and information necessarily translate into learning and deliberation. The purpose of deliberative dialogue is not to transform citizens into policy experts and have citizen engagement replace expert deliberation or input, especially around complex issues laden with technicalities and specialized knowledge. Also, in our experience, too much information can become overwhelming.

Citizens contribute in terms of other important policy-making considerations, primarily the clarification and prioritization of values. Engagement processes should provide a coherent view on how different pieces fit together and not necessarily communicate all available information (Harwood Group

1993). Hence, our objectives were to enable focused learning, in concert with values-based discernment and thoughtful consideration of the trade-offs that are inevitably embedded within different choices, to arrive at reasoned recommendations or advice on preferred directions or decisions. We also felt it critical to acknowledge and address the emotional dimensions of climate change, which required a different type of learning.

How We Responded

A participant handbook was used in each of the three deliberations. These primers provided information on issues from global and local perspectives, outlined key concepts of scientific uncertainty and risk, described deliberation and the role of values within this, and included a glossary of terms. In all three deliberations, the handbook was provided in advance and the majority of participants stated that they had read it before the discussion. Indeed, for AEEC, a poll taken during the discussions showed that 91 per cent of respondents had reviewed the participant guide in advance, providing a shared foundation for the deliberation.

Each of the handbooks had its own nuances, given the nature of the issue being explored and the deliberation format. For example, the Edmonton Panel handbook was a companion document to the highly technical Energy Transition Discussion Paper. It was designed to demystify climate change as well as provide the range of information noted previously. The WCC handbook introduced a social justice dimension to climate change. For AEEC, this primer oriented participants to their role, provided information designed to provide a common foundation of knowledge for all participants, and supplied detailed points about funding and regulation to consider for each discussion.

Scientific technical information was also provided in other formats. The WCC and Edmonton deliberations included informational presentations and panels. For example, in the WCC deliberation, a climate scientist made a presentation on climate change's probable impacts on southern Alberta. For the Edmonton Panel we invited experts from the city to go deeper into the discussion paper, followed by a carousel process in which participants rotated through small group discussions of the discussion paper's six goals. City staff served as resource people and participants were encouraged to identify what additional knowledge and information they needed to be able to move ahead with their deliberations. These presentations and panels gave participants the opportunity to ask questions and test out their own experiences/observations.

Given its duration, in the Edmonton Panel we were especially concerned with not overwhelming the participants with data, facts, and research. We deliberately did not front-load the sessions with reams of technical and scientific facts and figures. Instead we chose to introduce critical information iteratively and in response to participant requests through weekly reports that summarized previous session highlights, and previews of the coming sessions, with links to additional resources. In addition, during the opening session participants explored and shared their hopes and fears about different possible energy and climate change futures through the use of two scenarios (status quo and aggressive action). Encouraging participants to connect personally with this issue and share and listen to others' points of view helped to make climate change and energy transition more real for them.

It was also important to bring participant knowledge into the deliberation and provide opportunities for people to share and learn from lived experience. This was done, for example, by ensuring diversity of views through predetermining small group composition and allowing time for sharing of stories in both full and small group settings. In the WCC deliberation, this was taken a step further by designing a process that was sensitive to the participation of people from First Nations communities. This included requesting input from one First Nations participant into the process design and incorporating artwork from another participant on the backgrounder. The deliberation itself opened with a welcome and prayer, and time was provided for First Nations participants to share their experiences of how the climate was changing in the region and the impact this was having.

Strengths, Gaps, and Questions

While considerable time and resources were invested in the creation of the participant handbooks, overall they seemed to have been an underutilized learning and deliberation resource, especially in the Edmonton Panel. The reasons for this are not entirely clear to us, but one factor might be that we were often squeezing so much into each session that our references to the handbook might have seemed an afterthought rather than central to the participants' program of learning. At the end of each session, we included a preview of the next session with recommended reading from the handbook, but we might have provided additional prompts during the week to remind participants to review the material and contact us with questions and comments. However, we were balancing concern about potentially overtaxing them between sessions with a desire that they do some advance thinking on upcoming session topics. Overall, we think that well-crafted and balanced participant handbooks are a valuable aid to citizen learning and deliberation. However, to maximize their benefit for citizens and outcomes, they need to be explicitly woven into the deliberation process, and assessed and augmented as required.

We struggled with, and continue to wrestle with, questions about balance: How much learning and knowledge acquisition is enough for good deliberation among diverse participants? How much deliberation is enough for reasoned recommendations? What appropriate measures should we use to determine if participants are sufficiently informed to come to reasoned judgments on the issues at hand? These are all questions that, particularly for the Edmonton Panel, required continuous and iterative discussion with our partners. At one point, for example, the City asked to test panelists' knowledge and we reluctantly agreed to use keypads for this purpose. Our reluctance was due in part to our concern about what this "quiz" might signal to panelists if their knowledge was found wanting, and in part to the limitations of what binary, yes/no answers reveal about citizen knowledge. Fortunately, all panelists passed the quiz and there were no negative reactions to it! These processes served to build greater trust between the city and participants but also between the city partners and the project team. We continued to have different views on how much scientific and technical knowledge was needed for good deliberation. While acknowledging these differences, the ABCD team and its city partners came to better appreciate and respond to each other's respective perspectives and fears and to sharpen collaboration processes as a result (see chapter 6 for a more detailed discussion of the collaboration).

Enabling Citizen Deliberation

Challenge: Leverage the Unique Contributions of Citizens

A core dimension of citizen deliberation focuses on citizen values and value tension (internal and collective) in helping participants come to reasoned and ethical choices about policy options (Pidgeon et al. 2014). Citizens are invited to reflect on what values should guide government as it makes decisions and what value tensions need to be addressed. They are asked to apply those values to the issue(s) at hand, including thinking through the trade-offs that they are willing to make for the collective good. In most citizen deliberations, a discussion of values is crucial because the policy choices cannot or should not be made on technical or scientific grounds only.

How We Responded

Values work was integrated, to a greater or lesser extent in each of the three deliberations, to help ground participants' learning and reinforce their unique role as citizens in a democracy. We utilized adult education approaches, which stress the importance of experiential and hands-on learning, and principles for effective dialogue and deliberation, which recognize participants' competencies and learning needs (McCoy and Scully 2002; Schwartz 2002; Gastil 2014; Synapcity 2016). We worked to convey the idea that policy and action choices privilege different value choices, and we asked participants to be conscious and explicit about what value choices they were making as they deliberated on recommendations and came to decisions.

In both the Edmonton Panel and the WCC deliberations, an early presentation on values was scheduled that covered what values are, including how they differ from preferences and interests, and their place in family, community, and democratic life. This was done to orient participants to ways in which values could be integrated into their learning and deliberation on climate change. Then participants identified priority values that connected to the issues at hand for them. For example, in the Edmonton Panel, participants individually and then collectively, in small groups and then all together, identified priority values to guide their learning and deliberation, and discussed the meaning of each priority value to build shared understanding. Panelists used electronic keypads to select their top four priority values to serve as guideposts for their work together. This helped ensure that everyone's preferences were registered in determining the most important values and avoided having dominant and more powerful voices unduly influence decisions. The fourth session featured an expert's presentation on the psychology of climate change action and the alignment and misalignment of values with actions. This was designed to deepen panelists' understanding of the internal barriers and fears we face around climate change. We also invited participants to connect with their own experiences and to explore scenarios that depicted typical Edmontonians representing different lifestyles, values, and perspectives on climate change and energy. This was employed to help participants consider divergent and conflicting points of view, and the underlying values held by different Edmonton residents. The Edmonton Panel's final recommendations and report featured four value-driven principles and a core set of four values:

sustainability, equity, quality of life, and balancing individual freedom and the public good.

In AEEC, participants were tasked with considering a range of policy options and describing the conditions that would enable them to support a particular option. Participants were not asked to identify shared values to apply to these issues; instead each participant drew from her or his own personal values. For example, social justice was a value brought to bear on the discussion; the impact of a policy on low- and fixed-income homes was the most common issue raised in two of the four discussion areas. Other values underlying participant recommendations included cost-efficiency, transparency, accountability, and equity.

Strengths, Gaps, and Questions

The Edmonton Panel pre- and post-surveys and participant feedback at the sessions gave us some confidence that we had done a reasonable job of incorporating values into participants' learning, deliberation, and decision making. However, we heard from a minority of panelists that they were not convinced of the "value of values." They seemed to be unable to connect their values with the choices to be made and were looking for more information and research to help them to decide. This is not unusual in citizen deliberation, as many people are not used to thinking explicitly about values. With more time, we could have worked more with these individuals to better respond to their points of view, to understand why the values orientation did not resonate with them, and explore more deeply their unease. As well, a more explicit focus on the alignment between panelists' espoused values and their decision making around recommendations, including the emergent "Go faster, Go further" recommendation, might have resulted in a different order of thinking about the solutions required—beyond what the discussion paper provided. For example, for each recommendation, the report could have included an explanation of how participants aligned their position on the recommendation with their values and associated trade-offs.

In WCC, the values and concerns articulated in the morning were clustered into themes that became the key areas of discussion for the afternoon breakout groups. Given this, several of the themes included values as part of the cluster of ideas linked to that theme. One theme, social justice and responsibility, was expressed in values terms, and the recommendations continued to embed values (e.g., foster individual stewardship for development of the common good). For other themes and the recommendations on these, there was less explicit use of values. It is hard to move from the language of values into tangible recommendations. As citizens, we are not accustomed to doing this, and it is difficult to achieve in a one-day deliberation.

From our perspective, the early work on values, though not without its challenges, proved its worth. It helped to validate the importance of what citizens bring to the table in contrast to feeling that the deliberation was only about what was technically the best solution(s). For complex issues that require action by citizens (as well as society actors, including government and private sector agencies), aligning policy with values will have the most likelihood of success (Yankelovich 1991; Burall 2015). If the best technical solution rubs up against core values that are fundamental to the issue at hand, the solution may be resisted.

In our collective experience, it is important for citizen deliberation to go beyond the pronouncement of values important to the issue. While many of us share a bedrock of values, tensions can arise when we have to prioritize those values in considering the issue at hand. Understanding the public judgment citizens arrive at through deliberation is not complete without understanding how people have prioritized the critical values and the trade-offs they have made in doing so.

Challenge: Respect Participants' Diverse Life Experiences, Ideologies, Education, Expertise, and Ways of Knowing

The diversity of participants' life experiences captured in each of ABCD's deliberations, combined with the complex issues we asked participants to wrestle with, meant that it was particularly important that learning and deliberation materials, methods, and presenters were accessible for different needs, personalities, and circumstances. This is true of most deliberative dialogues that involve diverse participants. But layered on top of that was our challenge of ensuring that participants had a reasonable grasp of the key dimensions of energy transition and water and climate change. We worked hard to ensure that we were not triggering fear, intimidation, or avoidance.

How We Responded

In the Edmonton Panel and WCC deliberations we employed varied learning and deliberation methods to meet the diverse needs of the participants. Small group and plenary exercises were designed to respond and appeal to different adult learning preferences and aptitudes. While there was a strong cognitive orientation to the plenary learning sessions, the small group and some plenary exercises reflected experiential learning approaches that featured concrete experience, observation, reflection, thinking about concepts, and applying new knowledge (Schwartz 2002). These exercises included participants' use of photographs to elicit hopes, fears, and values; graphic presentations of group work to appeal to different ways of understanding; physical movement to activate different parts of participants' brains; and choices about discussion topics to allow participants to focus their energy on what was most relevant to them. While deliberation theorists often emphasize highly rational and cognitive aspects of deliberation (Habermas and Outhwaite 1996), our experience is that emotional and social dimensions of group deliberation are also very important and need to be attended to with equal care (Gastil 2014; Goleman 2006).

An important way of attending to participants' needs in all three projects was having teams of small group facilitators and note takers who had received prior training and orientation and whom we supported throughout the deliberation with detailed process guides, resource materials, and pre- and post-session briefings. The use of small group facilitators and note takers is a common practice in engagement work. But there are common challenges associated with the practice due to the sophistication of facilitation required for deliberation processes. Nonetheless, teams alerted us to challenging behaviours and situations so they could be addressed quickly before participants became distracted or stressed. This was critical to creating a safe and comfortable space for participants.

Strengths, Gaps, and Questions

As one might expect, the shorter the deliberation the harder it is to incorporate varied learning approaches. In the wcc deliberation we were able to use physical movement, small group and plenary work, visual methods such as the clustering activity, self-selection, and designated groups, but iterative learning processes were not a feature, as it was only a one-day session. For the AEEC deliberation, the two-hour time frame, made even shorter due to technology issues, limited the application of mixed methods. In most small group discussions, there was insufficient time for going beyond sharing participant views into a deliberative weighing of trade-offs and options, and there were few instances of facilitators having enough time to learn about participants' needs and alter the program accordingly. In the Edmonton Panel, however, we were able to incorporate a diversity of learning and deliberation methods to respond to the diversity of participants' needs and preferences. Still, we wonder if additional or different approaches might have been even more powerful in helping participants engage with the issues more deeply and holistically, and with a greater sense of ownership. For example, organizing field visits to Edmonton neighbourhoods or utility facilities, or allowing opportunities for more hands-on experiences might have contributed to other ways of thinking about energy transition and climate change in Edmonton and beyond.

Challenge: Be Attentive to Participant Ownership and Power Dynamics

We worked to design processes that did not privilege the already privileged (i.e., those who have higher education and are cognitively advanced and verbally articulate) and to address emerging power dynamics as they arose. We know that participants with higher education, more outgoing personalities, more experience speaking in groups, and more exposure to diverse situations are more likely to feel comfortable and confident about their participation (Hobbs 2013).

The context and purpose of a deliberation help to guide the level of ownership the project partners hope to engender in participants over the course of the dialogue. In some processes, it is quite appropriate for participants to deliberate and then walk away, having provided their best thinking to the hosts. Indeed, in policy processes, a common participant comment is that they are glad they don't have to actually make a decision given their new-found appreciation for how complex such policy/decision making is, and given the complexity of the issue and the different perspectives, interests, and values that have to be considered. In the WCC and the AEEC deliberations, the process was not designed to build long-term ownership of the results or move people to action. However, in the Edmonton panel, participants were encouraged and lightly supported to stay involved as their advice moved to City Council.

How We Responded

For those less comfortable speaking in large groups, their input and perspectives can be lost unless processes are designed to explicitly enable them to comfortably participate and contribute. To help achieve the active participation of all citizens in all three deliberations, we used a mix of small group and plenary processes. The small groups were facilitated to help participants live up to the ground rules of dialogue that they had created, one of which was "share the air time." We also varied the type of activities so that there were moments when different learning styles could shine; for example, those who prefer expressing ideas visually rather than through words, or in the case of the online deliberation, using typed comments rather than voice-based discussion.

For the in-person deliberations, we also incorporated time for individual reflection and followed this with sharing in groups of varying sizes. As we got to know people, we also assigned specific individuals to different groups to help balance the power dynamics, for example, putting all the dominant extroverts in one group, which freed up other groups from dealing with the dominant personalities.

For the Edmonton Panel, we provided structured and iterative processes for the deliberation, and redesigned on the spot in response to what was emerging in the discussion in order to help participants come to decisions on their recommendations. The better part of a day was scheduled for participants to identify their level of agreement for goals and activities, identify their trade-offs, and consider what values underpinned their choices. The use of keypads in the Edmonton Panel was particularly important in making a more level playing field for all participants and in ensuring transparency. Voting was anonymous and results were immediate, not filtered through an intermediary. The keypads were used throughout the panel process as icebreakers, to test knowledge, and to capture the pulse of the room; they were critical for the deliberation and decision-making phases, especially for voting on recommendations. For example, the recommendation that the city should "Go faster, Go further" to achieve greater carbon reductions was voted on in session five and then revised in the final session with a new vote, securing support from 63 per cent of participants. In thinking about the way in which this recommendation emerged and how participants navigated this, we draw a few conclusions. First, citizens felt empowered to bring their own reasoned and values-based recommendations into the deliberation. Second, participants and facilitators co-created a decision process around which recommendations to include and how to include them in their report (with 63 per cent agreement the "Go faster, Go further" recommendation fell short of the 75 per cent agreement required for inclusion). This process showed a sophisticated progression in their citizenship capacities to learn, deliberate, and decide.

In the WCC deliberation, a sense of ownership was built through the framing process in the morning, and then participants self-selected which of the resulting issue areas they wished to work on in the afternoon.

Participant-prepared reports also create a sense of ownership for both the deliberation process and the results. Of the three deliberations, only the Edmonton Panel incorporated this approach. It was a clear expectation from the beginning that the volunteer members of the panel would take a lead responsibility in overseeing the report writing, checking back with the whole group and drawing support from ABCD as needed. A core group of panelists presented the report to Executive Committee of Edmonton City Council in April 2013. The report informed the Implementation Strategy for Edmonton's Energy Transition Strategy, which was presented and accepted/approved by Council in 2015. The core group of citizen panelists were present both times. For the other two deliberations, a report was developed from note takers' and facilitators' notes of participant discussions.

AEEC sent a preliminary report to participants for their feedback before the final report was prepared and distributed back to them. AEEC made use of the findings in its engagement with the Government of Alberta in winter 2014.

Strengths, Gaps, and Questions

Despite our best efforts as designers and facilitators, we know that in time-limited deliberations we cannot eliminate all power imbalances among participants. So much of this happens at a subconscious level that we may not be fully aware it is going on (Choudhury 2015). When the power imbalance becomes visible, we can address it through design and facilitation processes. We also have to anticipate power dynamics and work to mitigate their effect, using techniques such as those described above, with the goal of ensuring that all voices are heard and respected (for more discussion on this issue, see chapter 5).

While we worked hard to prevent and mitigate power differentials and to encourage ownership, we are left with some questions. We wonder if providing opportunities for social learning and interaction (e.g., informal dinners or pub nights) might have generated greater group trust, rapport, energy, and ownership of issues (Collins and Ison 2009). Unfortunately, budget and time constraints precluded these options. Nonetheless, we might have thought creatively about ways to encourage more social learning during and outside the formal sessions. For example, we might have offered additional learning and discussion sessions where participants could meet with resource people to discuss topics of particular interest or concern. In addition, we wonder if all three projects could have provided better opportunities for participants to work through what it means to be an effective citizen in relation to energy transition and climate change. Questions such as: "What does being an active citizen mean? What are the ways to effectively engage in policy? And how do I participate in civic life?" are not part of everyday discourse, and our political and media culture does little to encourage such conversation (Synapcity 2016).

Given the time constraints of each project, it was difficult to integrate learning and discussions about what it means to be an effective and informed citizen into the deliberation agendas.

Concluding Reflections and Questions

Although we have been practitioners of public deliberation for many years, each new deliberation continues to challenge us, deepening our learning, testing our design and facilitation skills and challenging our thinking. This was especially true of deliberations focused on climate change and energy that occurred in the complex context of a fossil fuel-producing jurisdiction. Our ABCD experience has elicited new understanding of the place and importance of issue framing, participant diversity, working with partners, and putting participant values at the centre of deliberations. It has also heightened the need for personal reflexivity about our roles as designers and facilitators.

During these three deliberations, our personal convictions about the necessity and urgency of decisive action on climate change commingled with a responsibility to design and facilitate successful deliberative processes that would contribute to policy, citizen action, and citizen capacity building. This work was complex and challenging. The issue of climate change can be daunting, disempowering, or discouraging for all of us. Connecting with these emotions was an essential piece of being an effective member of the deliberation team. Participants went through equally, if not more, intense experiences, and as lead facilitators we needed to pay attention to our inner uncertainties before inviting others to do so. Failure to do so, we feared, might have meant that we were not fully present to what was happening in the room—virtual or real. Being aware of our vulnerabilities and triggers, and then working to support each other (in the case of the Edmonton Panel) to ensure we were attuned to participants' emotional, social, and cognitive needs, took regular reflection and check-ins. We kept asking whether our processes enabled citizens to do their best work and whether they were furthering the best deliberations possible in the real world context within which we were all working.

In addressing these questions we played multiple roles, including those of process experts, partners, collaborators, facilitators, and citizens. We juggled these roles, working to hold our ground on best practices, while also reorienting agendas at the last minute to provide time to accommodate partners' or researchers' objectives. On reflection, managing multiple roles and responsibilities is critical to reflexive learning and growth. One lesson we took from our experience is that our growth and development as facilitators will be short-changed if we fail to take the time and space during and after these intense projects for reflective work. Knowing how challenging it is for facilitators to find the time and space for this reflection, we wonder what supporting roles deliberation academics and "pracademics" could play in that regard.

In this chapter, we highlighted the opportunities and constraints of framing issues for deliberation. We believe that deliberations connected to immediate policy opportunities have value and can increase the impact of deliberations, but at the same time, using a systems frame for climate change deliberations could have greater impact on long-term goals. Wrestling with a more complex framing of climate change might empower citizens to come to a more nuanced understanding of the issues. It might also help them to better prioritize issues of greatest importance to them, which could cultivate their interest and capacity for ongoing involvement (see chapter 8). We are left wondering what an agenda would look like that empowers participants to determine if their discussions should focus on the policy opportunity or the issues most important to participants, and what would enable the convener to support such a responsive format.

If institutions and policy makers were to take a longer-term view by engaging with the complexity of climate change now, further resiliency could perhaps be fostered and capacity increased for unanticipated challenges to come. This could be accomplished by institutionalizing public engagement as a regular input into policy formation and other decision making and community action. Sustained opportunities for involvement would provide the time and depth for the public to engage, reflect, and deliberate over a longer term, a requirement for adequately exploring the complexity of climate change issues.

All key decisions for the ABCD deliberations were made in partnership with convenors, researchers, and practitioners. We have observed through ABCD and other projects that it usually takes direct exposure for people new to the field to understand what public deliberation actually is and how much citizens are capable of, and to trust in the unfolding process. It is also true that practitioners are responsible for entering into the mindset and contexts of policy makers and academics in order to understand and respond to their needs and perspectives. Building shared understandings and trust takes time and should not be underestimated. A strong foundation can enable the team to be flexible and to respond effectively to the needs of citizens, while ensuring that all partners feel their needs are still being met through any necessary design changes (see chapter 6).

Ultimately, we think that the needs of participants are the most important drivers in designing climate change deliberations. What do participants need in order to do their best learning, discussing, and deciding? We believe that participants' work must be rooted in their values, because values are an essential piece of the policy puzzle that the public brings to the table. Yet, in general, citizens are not practised in reflecting on the role of values in public policy choices, so designers should not overestimate people's ability to talk about and think that way. It might be necessary to support participants in seeing how their values are implicit in the decisions they make: work that is particularly challenging during climate deliberations because there can be both emotional and cognitive dissonance between participants' world views and their actions. This challenge becomes even more difficult in the face of time limitations.

Understanding the advantages of values-based deliberation is also relevant for decision makers and their supporting institutions (Nabatchi et al., 2012). The ABCD recommendations were based more or less explicitly on values, but sometimes the values themselves were a key outcome, such as with the Edmonton Panel. We also wonder about how institutions interpret and translate those values. Further attention to this area would benefit climate change and other public deliberations. Our experiences have also whetted our appetite to develop innovative approaches to embedding values within citizen learning and deliberation, and we invite practitioners, policy makers, and academics to help us do so.

References

Abelson, Julia, and François-Pierre Gauvin. 2006. *Assessing the Impacts of Public Involvement: Concepts, Evidence and Policy Implications.* http://www.cprn.org/documents/42669_fr.pdf.

Barisione, Mauro. 2012. "Framing a Deliberation. Deliberative Democracy and the Challenge of Framing Processes." *Journal of Public Deliberation* 8(1): art. 2. http://www.publicdeliberation.net/cgi/viewcontent.cgi?article=1176&context=jpd.

Burall, Simon. 2015. *Room for a View: Democracy as a Deliberative System.* London: Involve. http://www.involve.org.uk/wp-content/uploads/2015/10/Room-for-a-View.pdf.

Choudhury, Shakil. 2015. *Deep Diversity: Overcoming Us vs. Them,* Toronto: Between the Lines.

Collins, Kevin, and Ray Ison. 2009. "Jumping off Arnstein's Ladder: Social Learning as a New Policy Paradigm for Climate Change Adaptation." *Environmental Policy and Governance* 19(6): 358–73.

CPEECC (Citizens' Panel on Edmonton's Energy and Climate Challenges)
2013. *Citizens' Panel on Edmonton's Energy and Climate Challenges Report.*
https://www.edmonton.ca/city_government/documents/PDF/CitizensPanel-
EnergyClimateChallenge.pdf.

Frank, Shannon. 2015. Telephone Interview with Lorelei Hanson, July 19.

Gastil, John. 2014. *Democracy in Small Groups: Participation, Decision Making &*
Communication, 2nd ed. State College, PA: Efficacy Press.

Goleman, Daniel. 2006. *Emotional Intelligence: Why It Matters More Than IQ.* New
York: Bantam.

———. 2013. Focus: The Hidden Driver of Excellence. New York: HarperCollins.

Gutmann, Amy, and Dennis Thompson. 2004. *Why Deliberative Democracy?*
Princeton, NJ: Princeton University Press.

Habermas, Jürgen. 1996. *Between Facts and Norms.* Translated by William Rehg.
Oxford: Polity Press.

Habermas, Jürgen, and William Outhwaite. 1996. *The Habermas Reader.* Cambridge,
UK: Polity Press.

Harwood Group (now Harwood Institute). 1993. "Meaningful Chaos: How People
Form Relationships with Public Concerns." Report Prepared for the Kettering
Institute. https://www.dropbox.com/s/xb86hr906m5r8a2/Meaningful%20Chaos.
pdf?dl=0.

Hobbs, Lindsay. 2013. "Participant Experiences of Change in a Deliberation Setting."
Alberta Climate Dialogue Working Paper. http://www.albertaclimatedialogue.
ca/participant-experiences-of-change-in-a-deliberative-setting/participant-
experiences-of-change-in-a-deliberative-setting/.

Kettering Foundation. 2011. "Naming and Framing Difficult Issues to Make Sound
Decisions." Kettering Foundation Report. https://www.kettering.org/sites/default/
files/product-downloads/Naming_Framing_2011-.pdf.

Levin, Kelly, Benjamin Cashore, Steven Bernstein, and Graeme Auld. 2012.
"Overcoming the Tragedy of Super Wicked Problems: Constraining Our Future
Selves to Ameliorate Global Climate Change." *Policy Sciences* 45(2): 123–52.

Lukensmeyer, Carolyn. 2012. *Bringing Citizens' Voices to the Table: A Guide for Public*
Managers. San Francisco: Jossey-Bass.

MacKinnon, Mary Pat, Jacquie Dale, and Deborah Schrader. 2014. "Looking under
the Hood of Citizen Engagement: The Citizens' Panel on Edmonton's Energy
and Climate Challenges, ABCD." Alberta Climate Dialogue Working Paper.
albertaclimatedialogue.ca/cl4_looking-under-the-hood.

McCoy, Martha L., and Patrick L. Scully. 2002. "Deliberative Dialogue to Expand Civic
Engagement: What Kind of Talk Does Democracy Need?" *National Civic Review*
91(2): 117–35.

Nabatchi, Tina, John Gastil, G. Michael Weiksner, and Matt Leighninger, eds. 2012. *Democracy in Motion: Evaluating the Practice and Impact of Deliberative Civic Engagement*. Oxford: Oxford University Press.

Nabatchi, Tina, and Matt Leighninger. 2015. *Public Participation for 21st Century Democracy*. San Francisco: Jossey-Bass.

Owen, Harrison. 1997. *Open Space Technology: A User's Guide*. San Francisco: Berrett-Koehler.

Pidgeon, Nick, Christina Demski, Catherine Butler, Karen Parkhill, and Alexa Spence. 2014. "Creating a National Citizen Engagement Process for Energy Policy." *Proceedings of the National Academy of Sciences of the United States of America* 111(4): 13606–13. doi:10.1073/pnas.131751211.

Pike, Cara, Bob Doppelt and Meredith Herr, 2010. "Climate Communications and Behavior Change: A Guide for Practitioners" Climate Change Initiative, https://scholarsbank.uoregon.edu/xmlui/handle/1794/10708.

Prikken, Ingrid, Simon Burall, and Michael Kattirtzi. 2011. "The Use of Public Engagement in Tackling Climate Change." London: Involve. http://www.involve.org.uk//wp-content/uploads/2012/01/The-use-of-public-engagament-in-tackling-climate-change.pdf.

Rose, Jonathan. 2007. "The Ontario Citizens' Assembly on Electoral Reform." *Canadian Parliamentary Review* 30(3): 9–16.

Schwartz, Roger. 2002. *The Skilled Facilitator: A Comprehensive Resources for Facilitators, Managers, Trainers and Coaches*. San Francisco: Jossey-Bass.

Smith, Mark K. 1997, 2004. "Carl Rogers and Informal Education." *Encyclopaedia of Informal Education*. Last updated May 29, 2012. www.infed.org/thinkers/et-rogers.htm.

Synapcity. 2016. "About Civics Bootcamp." http://synapcity.stiff.ca/workshop/civic-boot-camp/.

UN (United Nations). 1992. Rio Declaration on Environment and Development. http://www.unep.org/documents.multilingual/default.asp?documentid=78&articleid=1163.

——— 2015. Framework Convention on Climate Change. December 12. http://unfccc.int/documentation/documents/advanced_search/items/6911.php?priref=600008831.

Warren, Mark E., and Hillary Pearse, eds. 2007. *Designing Deliberative Democracy: The British Columbia Citizens' Assembly*. Cambridge: Cambridge University Press.

Woodruff, Paul. 2005. *First Democracy: The Challenge of an Ancient Idea*. Oxford: Oxford University Press.

Yankelovich, Daniel. 1991. *Coming to Public Judgment: Making Democracy Work in a Complex World (Frank W. Abrams Lectures)*. Syracuse, NY: Syracuse University Press.

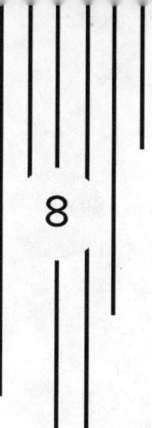

8

Climate Change, Social Change, and Systems Change

David Kahane

Alberta Climate Dialogue (ABCD) came together because a group of leading researchers and practitioners of deliberative democracy wanted to make a difference in the realm of responses to climate change using the tools of public deliberation. I led the development of ABCD, was its Principal Investigator and Project Lead, and was involved in planning and researching each of its four deliberations. Our team believed that deliberation could make a difference in policy responses to climate change in Alberta, and we thought of public deliberation as a component of systemic change. Now is the time to ask and attempt to answer the questions of: What difference did we make, and what can we learn from our efforts at change? And what did we learn about the strengths and limitations of deliberative democracy in addressing a complex systemic problem like climate change?

In what follows, I unpack the character of wicked or super-wicked systemic problems, focusing on the example of climate change. I next describe eight stories of social change and impact—told within the deliberative democracy community—and use these to look at ABCD's impact. I suggest that neither the field of deliberative democracy nor ABCD has been highly focused on whole systems and I explore why this might be the case. Finally, I examine how deliberative democracy can use insights from the fields of systems thinking, user-centred design, and systemic design to better address complex challenges like climate change.

Climate Change as a Complex Systemic Challenge

Deliberative democrats believe that engaging diverse citizens directly in problem solving and policy development can improve the responsiveness, accountability, and effectiveness of government, and build effective responses to our toughest challenges. Deliberative democratic exercises always take place in particular contexts and on limited scales, and yet often aspire to make a difference within large, complex systems. When representative groups of citizens are convened to help governments develop policy on climate change, for example, they work within a particular issue frame or set of frames, focus on a particular jurisdiction or set of jurisdictions, and seek action in particular places or around particular policies. Yet we know that climate change is a global challenge cutting across every jurisdiction, can be approached through a wide diversity of frames (see chapter 5), and touches just about every area of human activity and politics.

The dominant frame for climate responses in Canada treats climate responsibility as congruent with an advanced capitalist economy and with economic growth. The interventions required for progress, according to this perspective, have to do with skilful market transformation: using education, social marketing, subsidies, taxes, and regulations to shift market behaviours of individuals and firms so that we transition quickly to a prosperous low carbon economy. There also is recognition in at least some quarters of the public, civil society, and government of the need to increase community resilience in the face of climate impacts that will become more severe.

The challenges around climate change are serious even if you believe that mitigation and adaptation are possible through reforms to existing social, economic, and political structures. If you believe that climate change is a symptom of deeper pathologies in social, economic, and political systems that require deep transformation or revolution, as Klein does (2014), questions of boundaries and framing, and about influence and interaction across scales, sectors, and time, become thornier still. So one's underlying theory of social change matters: How will your intervention interact with other forces to bring about the shifts you seek?

The difficulty of placing a particular public deliberation exercise within a persuasive story of social change is multiplied when we appreciate climate change as a wicked or super-wicked problem. With wicked problems, issues are defined differently by different stakeholders; understandings of the problem evolve; there is no right solution and no learning through trial and error;

and the problem cuts across systems of governance as well as being viewed by many as a symptom of other problems (Rittel and Webber 1973). It has been suggested that climate change is in fact a super-wicked problem, with all of the features just cited and more: time is running out; those trying to solve the problem are also causing it; the solution arguably requires a central authority but central authorities are non-existent or weak; and populations as well as policy makers irrationally discount the future costs of current behaviours (Levin et al. 2012; see introduction).

Interventions around climate change are embedded in non-linear systems. Systems are often defined in contrast to more linear models that isolate particular elements of causation from the intricate temporalities and feedbacks that shape outcomes. A non-systemic deliberative democratic description of the work ABCD did with the City of Edmonton, for example, might start with a simple causal chain: citizens will deliberate, they will generate a report, the report goes to City Council, Council makes a decision, and this policy changes the state of affairs in the world. And still thinking linearly, one could start accounting for other forces and players, building a model that situates a deliberative intervention within a complicated set of dynamics.

Systems thinkers urge that we approach things in more ecological terms. Changes in ecosystems do not happen in straight lines but through negative feedback loops (where a perturbation feeds into systems that bring things back into balance, as when a healthy body deals with fluctuations in body temperature) and positive ones (where a stimulus causes changes that increase its power, as with global warming melting permafrost and releasing methane that increases warming). These feedback loops have different time lags, which—combined with how any given system nests within other systems—makes the change caused by any particular intervention intricately complex.

Where a linear understanding of the complications of linking citizen deliberation to political outcomes might be analogized to changing a setting within an intricate machine with many moving parts, a systems understanding might instead picture a novel event in a forest ecosystem, the impacts of which emerge through webs of complex interdependence. Such a system is more than the sum of its parts, and the behaviour of the various parts arises from the structure of the whole.

> An implication of interdependence is that actions have effects other than
> those intended. Since everyone always sees and acts locally, there is no

reason to expect that an aggregation of incremental improvements will improve the greater whole. Systemics exposes an assumption we have organised our societies around. This is the assumption that knowledge and action are both furthered when we divide them into smaller pieces over and over again.[1] (Ryan 2014, 3)

Systems theory plainly applies to climate change, one of the most complex systems problems humanity has ever faced (van der Lans 2014). In dominant climate change approaches, interventions tend to focus on altering parameters that may be superficial in terms of systems dynamics. For example, developing cleaner technologies for oil sands extraction may reduce emissions but not touch deeper dynamics of a capitalist, consumerist system premised on cheap sources of carbon-based energy. Indeed, perceived leverage points may even push in the wrong direction: cleaner oil sands technologies could perpetuate the illusion that oil sands can be a sustainable form of energy in the face of climate change (see Easterbrook 2011).

Deliberative Democracy and Theories of Change

What understandings of social change tend to be implicit in work that travels under the banner of deliberative democracy, and in conversations that take place in gatherings of practitioners and researchers in the field? To what extent are these understandings adequate to deep social complexity, wicked problems, and the systems dynamics sketched above?

Let me start with a personal observation based on my experience over the last decade at gatherings where researchers and practitioners of public deliberation assemble to learn new practices and reflect on the state of the field: gatherings like the Canadian Conference for Dialogue and Deliberation (C2D2), the National Coalition for Dialogue and Deliberation (NCDD), and the International Association for Public Participation (IAP2). At meetings like these and in the field more generally, I've encountered relatively little sustained, collective thinking about how deliberative democracy contributes to social change or transformation. In these contexts, practitioners tend rely on a particular kind of story about their work:

> *It tends to be a reformist story*: citizen deliberation can be articulated with established decision-making structures to influence outcomes, while also changing experiences of citizenship and addressing

social injustice (e.g., Bohman 1996; Gutmann and Thompson 1996; Lukensmeyer 2012).

It tends to be an aggregative story: the remedy to problems with liberal representative democracy is more and more deliberation involving more and more people in more and more places, thus building capacity, linkages, and infrastructures over time (e.g., Nabatchi and Leighninger 2015).

It tends to be a "liberal" story: it focuses on the positive effects that the careful exercise of deliberation can have on collective decision making and action (e.g., Gastil and Levine 2005). I would contrast this with a "radical" analysis that would focus on how social stasis and change are explained by the unintended, systemically mediated effects of individual actions—that is, by structures and mechanisms of power that underlie and constrain our individual and collective reasoning and decisions (e.g., Young 2005). This is a spectrum rather than a binary, and deliberative democrats tend to sit at the liberal end.

This dominant story is a positive one: gathering individuals to deliberate on common projects builds individual and community capacity, and can support or push organizations and governments to better meet human needs. While I share some of this optimism, it is worth considering the negative potentials of deliberation and the ways in which public engagement can: disempower participants and reduce their agency; reinforce exclusions and hierarchies; be manipulated; build capacities that are used for corrupt or negative ends; or be used to produce reports and recommendations that are never taken up or implemented (Gaventa and Barrett 2010; C. Lee 2015; Johnson 2015).

In what follows, I tease out eight stories of social change from deliberative literature, practices, and conversations. These change stories are not neatly bounded or separate; practitioners, organizations, and theorists combine them in diverse ways. All eight describe how citizen involvement can bring about social and political change.

The first story of social change involves legal empowerment. Deliberative bodies are authorized by governments to make binding decisions or decisions that will be voted on directly by the public. Prominent examples of this include participatory budgeting, as well as citizens' assemblies on electoral reform in British Columbia (2004) and Ontario (2006).

The second story highlights a connection to government. Deliberative processes are formally linked to legislative processes so that outcomes shape decisions or increase responsiveness. Here, citizen voices and wisdom directly influence policy development, decision makers commit to taking recommendations seriously, and participants in deliberation often act as advocates for their recommendations (Lukensmeyer 2012; see Gaventa and Barrett 2010, 59, on the need for accompanying civil society mobilization).

The third story focuses on lobbying. Civil society organizations bolster their campaigns for changes to state behaviour by holding deliberative processes. Such approaches can blend citizen deliberation with multi-stakeholder processes to build influential coalitions of organizations outside government to push for state action.

A fourth story of how deliberative processes can have an impact focuses not on organizations but on individual citizens activated through public mobilization. Some deliberative strategies emphasize "critical mass"—getting many members of a community involved in dialogue and deliberation to build broad public pressure for government action. Other deliberative mobilization strategies emphasize building public confidence and constituencies for political change by communicating and legitimating the process and results of a deliberative exercise to publics not directly involved (Cutler et al. 2007).

The fifth story is about deliberative capacity. Experience with deliberative processes changes how governments, civil society organizations, grassroots communities, and deliberation practitioners engage with publics in the course of their work. A number of overlapping discourses and literatures fall under this heading:

- A broadly embraced discourse of capacity building as part of public involvement: practitioners from outside a convening organization or government support a deliberative process and at the same time seek to increase the organization's ability to understand, plan, and deliver future engagement processes (Lukensmeyer 2012).

- A more specialized discourse and literature on institutionalizing or embedding deliberation in how government and other organizations operate, so that rather than engagement processes being ad hoc and one-off, they become part of standard processes of decision making, or legally mandated as a right to participate (e.g., Gaventa 2006).

- Scholarly literature on deliberative systems and, more recently, deliberative infrastructure: rather than evaluating particular deliberation processes in isolation, these approaches show connections between diverse spaces of deliberation, including legislative bodies, mini-publics, civil society, media, and online spaces. These approaches explore how a political system can achieve deliberative and democratic goods as an integrated whole (Fagotto and Fung 2009; Lukensmeyer 2012; Mansbridge et al. 2012; Nabatchi and Leighninger 2015).

The sixth story concerns community empowerment. Grassroots deliberative work increases the ability of communities to act on their own and solve their own problems. This can involve building abilities for collective action, fostering new forms of participation, and deepening networks and solidarities (Gaventa and Barrett 2010, 27–32).

The seventh story focuses on including the excluded. Running through many of the above stories is the goal of lifting up the voices and increasing the political influence of marginalized and oppressed groups (Williams 1998).

The eighth story is about changing participants. Deliberative democratic exercises increase civic and political knowledge, trust in government, sense of empowerment and agency, and propensity to participate in civic life (Fung 2003, 350).

ABCD's Change Stories and Impacts

There are so many intervening variables in political processes that it is extremely difficult to reliably assess the impact of citizen deliberation exercises on policy and decision making (Delli Carpini, Cook, and Jacobs 2004; Mutz 2008; Ryfe 2005). Nevertheless, through its deliberation and research work, ABCD tried to assess some of those impacts. What follows is an overview of how ABCD's work played out across the eight stories of deliberative impact described above.

The first story, legal empowerment, was never part of ABCD's plans. Outside of participatory budgeting in municipal contexts, legal empowerment for deliberative exercises takes tremendous boldness on the part of political leaders and parties; this is rare in North America. I have never heard of legal empowerment of deliberative processes in connection with climate policy, and it likely would have been a non-starter in our context given the political sensitivity of the issue, the desire of political elites to maintain control of policy, and the dominance of expert discourses in decision making.

Connection to government was the core change story for much of ABCD's work. We decided early on to focus on partnering with municipal governments in Alberta to convene citizens for stronger climate action. The partnership that emerged was with the City of Edmonton: a Citizens' Panel on Edmonton's Energy and Climate Challenges (Edmonton Panel) was developed with the Office of the Environment at the city and the Centre for Public Involvement. The panel brought together fifty-six citizens for six days of deliberation; it yielded a set of recommendations submitted to City Administration and the Executive Committee of City Council, and that fed into an Energy Transition Strategy passed unanimously by City Council in 2015.

How did the Edmonton Panel influence city decisions? It's hard to assess how much the deliberation process and its recommendations shaped the content of policy. First, the terms of the deliberation and the policy were importantly set by a detailed discussion paper (Pembina Institute and HB Lanarc 2012), such that citizens weren't involved at a stage where they could influence the particular range of greenhouse gas (GHG) reduction mechanisms being considered by the city, or how climate change was framed. There was openness to citizens bringing their own frames and ideas into the deliberative conversation, but many parameters were already set. In the end, the Panel's recommendations supported the City's prior framing of energy transition as reflected in the discussion paper.

Second, the panel was asked to choose between three energy pathways for Edmonton—"business as usual," "reduced carbon," and "low carbon"—then to make recommendations on measures associated with their chosen pathway. The existing orientation of City Administration was toward the low carbon pathway, and citizen panelists affirmed this by a majority of 94 per cent. Moreover, while panelists offered caveats around particular city actions to achieve the low carbon pathway, by and large they supported the implementation measures advocated in the discussion paper. So here, as with framing, strong alignment between desired outcomes makes it hard to separate out the Edmonton Panel's influence.

Third, there was a long period of policy development between the Edmonton Panel's final report (March 2013) and the drafting of the Energy Transition Strategy (2014–15), with extensive further expert and stakeholder input and many shifts in framing and analysis of issues; this makes it hard to trace the influence of finer-grained panel recommendations. With that said, the Panel did have some of its specific recommendations reflected in the Energy Transition Strategy, especially around principles and values to guide city climate action, and advice on how to communicate with citizens (see Alberta Climate Dialogue and Centre for Public Involvement 2015).

What of the influence of the Edmonton Panel on political decision making? The evidence we have comes from interviews and public statements by civil servants and elected officials. One aspect of political decision concerned City Administration's level of ambition in formulating the strategy and their willingness to bring it to Council. Speaking after passage of the Energy Transition Strategy, the city manager most closely connected with the development of the panel, and a lead author of the Energy Transition Strategy, remarked:

> The work of the Citizens' Panel was really critical to the overall success of Edmonton's Community Energy Transition Strategy. Without it . . . there would have been a gaping hole. I don't think you can bring an effort like this to a council without some level of assurance or support from citizens. . . . Without that type of work, that quality work, you're just not ready to go to Council or else you should expect some big trouble. (Andrais 2015)

Upon passage of the Energy Transition Strategy, councillors and the mayor spoke about how the panel increased their willingness to support the strategy. In the mayor's words, "I think the Citizen Panel gave confidence to council that a representative group of citizens armed with the right information would come to the same conclusion that our Administration's recommending, which is that we should take action" (Iveson 2015). A city councillor said:

> I think the panel's influence was quite profound in the end because I think it did give everybody comfort that we were not out of line with what a group of citizens of this city coming together and deliberating were going to come to in terms of their understanding and their decision and their beliefs about how we should move forward as a city. . . . So, I think it was reassuring to know that what we suspected was there in terms of public support was actually there and to be able to test that. And to be able to know that if people really have a chance to look at this and weigh the options and understand that there's trade-offs, that these are the answers that they came up with. (Henderson 2015)

The City had invested heavily in the panel, and it provided useful rhetoric in favour of a strategy both the mayor and Council supported, so these significant positive statements need to be interpreted in that context.

The third change story is that deliberation hosted by civil society organizations can support their lobbying. Following the success of the Edmonton deliberation, ABCD held a funding competition for project members who wished to develop other kinds of citizen involvement processes. One of the successful proposals was for a set of two-hour, province-wide virtual deliberations by telephone and online, intended to support the lobbying efforts of the Alberta Energy Efficiency Alliance (AEEA), an environmental NGO. Jesse Row, AEEA's lead, had a strategic intention in convening the deliberation: he hoped to gather evidence that the public, given the chance to deliberate, supported the AEEA's advocacy for regulation of energy efficiency standards and greater provincial funding for energy efficiency programs; he also wanted information about the shape of public views (Row 2015). He later said that his ability to point to some particular voting numbers from a post-deliberation survey, alongside opinion polling AEEA did that was consistent with deliberation results, was a powerful argument in lobbying the Alberta government and others (2015). The piece of policy that the lobbying was meant to influence was never announced due to a change in government in 2015.

The fourth change story is one of public mobilization, which, as noted earlier, can be broken down into critical mass strategies that seek to involve a broad public, and strategies that rest on informing the public by communicating broadly about a deliberative process that involves relatively few participants. Critical mass approaches wove in and out of ABCD planning and discussions, including an unsuccessful proposal to the City of Edmonton for a citizens' panel process that would have involved civil society groups in supporting the work of the Edmonton Panel, and proposed having panelists and these organizations work together to convene further community-based conversations. In another example, the 2012 Edmonton City-Wide Food and Urban Agriculture Citizen Panel successfully mobilized citizens. Fifty-eight citizens deliberated over a six-week period and made recommendations to City Council; their top recommendation opposed the development of urban farmland in the city's northeast, alongside a host of other recommendations. The Food and Urban Agriculture panel drew public attention to these issues, and many participants in the panel participated in hearings before City Council and have remained active in the local food movement. In the end, City Council approved development of the northeast farmlands, to the great disappointment of civil society organizations, some panelists, and many involved citizens. Other panel recommendations to the city may have carried more weight, though here we run into the problem, noted above, of assessing impacts given many intervening variables.

What about building deliberative capacity, the fifth change story about impacts? The ABCD project team talked a lot about capacity building. We offered workshops that introduced civil servants, elected officials, and others to deliberative democratic methods. Moreover, in developing particular projects we sought to develop capacities in facilitators and note takers (see chapter 7) and hoped that the organizations we worked with would become more adept and more supportive of innovative citizen engagement. The impact of these efforts is hard to trace and most likely modest.

The sixth change story is one of community empowerment, which had only a faint echo in ABCD's deliberation projects. While the first iteration of our proposal to the City of Edmonton did envision community projects, the number of citizens involved was always going to be small, and this version of the deliberation was, in any case, rejected. The Food and Urban Agriculture Citizen Panel and surrounding activities, in which ABCD was more lightly involved, did connect with many forms of community ferment and action.

When it comes to including the excluded, the seventh change story, we can start by noting that ABCD as a project team had rough gender balance but was mainly white and class privileged. We worked repeatedly to increase representation of marginalized and oppressed groups, especially Indigenous people, in ABCD and our planning processes, with limited success (see chapters 4 and 6). The reasons are manifold: the whiteness of academia as well as the citizen involvement profession; the real and perceived irrelevance of research projects like ABCD (and its deliberation processes) to the pressing concerns of marginalized and oppressed communities; and the limited skills and networks of many of our Alberta-based members when it came to connecting with non-white, non-privileged groups and representatives. We had more success including diverse participants in our deliberations: the Edmonton Panel was in many ways demographically representative of the city, and the Oldman Watershed deliberation included a number of Indigenous participants. It is less clear that including individuals from marginalized and oppressed groups in deliberative spaces meaningfully increases the political influence, or addresses the marginalization and oppression of the communities from which they come (Gaventa and Barrett 2010, 44–46; von Lieres and Kahane 2007).

Finally, what of the eighth story, changing participants? ABCD invested heavily in survey research to measure the impacts of the deliberations on citizens. For example, for the Edmonton Panel we gathered data at the time of recruitment, and before, several times during, and after the deliberation. Our findings were inconsistent; they seem to confirm the view that research "provides a good deal of indirect support for the democratic potential of deliberation but also suggests that this potential is highly context dependent and rife with opportunities for going awry. Research explicitly devoted to the political consequences of deliberation, though relatively sparse, leads to a similar conclusion" (Delli Carpini, Cook, and Jacobs 2004, 328; see also Mutz 2008). We did document shifts in opinion: for example, participants in the Edmonton Panel had a greater confidence in their ability to affect what government does at the end of deliberation than before it started[2]; there was a rise in self-reported measures of participant knowledge about climate change, around what climate change is, what energy vulnerability is, ways to reduce Edmonton's GHG emissions, and more (see Hobbs 2013). Longitudinal evidence of change or the persistence of change was ambiguous in the survey data. While some panelists clearly were moved and influenced by their experience (this comes across, for example, in interviews with participants who remained strongly involved in the politics of municipal energy transition after the panel) (Hannah and MacLellan 2015), we can't claim that there were widespread or significant changes across the fifty-six participants.

It is worth reflecting on the fact that all four of the deliberations in which ABCD was involved were "mini-publics": spaces of deliberation designed by professionals into which a relatively small, deliberately recruited group of citizens are invited by conveners to learn about an issue, exchange perspectives, and make recommendations. As comes out in the discussion of this chapter, mini-publics have strengths but also weaknesses when it comes to understanding and supporting deep and systemic change. The fact that ABCD worked with this model speaks to the range of deliberation expertise in ABCD and how the project was able to draw on this under pressure, the needs and desires of the partners with whom we became involved, and a certain path-dependence that came out of early choices in framing ABCD's work (see chapter 6).

Impact on Deliberative Democracy

I have focused on particular deliberation projects in looking at these eight change stories in ABCD. What of changes seeded by the project as a whole? The project had its inception at a large meeting of researchers and practitioners of deliberative democracy asking, "How do we advance the field?" What was

ABCD's role in advancing the field, especially as it relates to the challenge of climate change?

In addition to holding deliberations, we in ABCD did a great deal of research, convening, and outreach to advance the project and the field. We hosted many workshops for different audiences on deliberation and climate change; shared our project's learning through social media, web platforms, and scholarly research; and built strong relationships within and beyond our team. These activities had ripples: relationships and networks were built among members of ABCD and with partners and others; thousands of conversations about public deliberation and climate change took place with civil servants, elected officials, participation professionals, NGO members, and fellow citizens; graduate students built their capacities through ABCD research assistantships; ABCD materials were taught in a number of university courses; there were numerous presentations to academics, practitioners, and civil servants; and team members produced dozens of publications, blog posts, and working papers. Our learning and expertise have fed into other deliberation and change projects like the Climate Justice Project's Conversation on Climate Justice in British Columbia (M. Lee 2015) and the Energy Futures Lab (www.energyfutureslab.com) in Alberta. The project director and others continue to be drawn into government conversations about public involvement on climate change at the provincial and federal level.

The influence of all of this is hard to trace. Ultimately, to make claims about the difference ABCD made through activities like these, one has to reference some contestable account of how change happens, and what holds the status quo in place—which returns us to questions of systems thinking and systems change.

Why Aren't Deliberative Democrats More Focused on Systems Change?

The Case of ABCD

As already discussed, our predominant story of social change in ABCD was supporting better climate responses by convening deliberations with governments to inform policy development. Yet, looking back on seven years of collaboration, this story of change raises crucial questions.

First, what conditions would need to be in place for our deliberative partnerships to shape policy in the most productive and progressive ways? For example, what other forces were in play in government, in political and democratic activities already going on in Alberta (Chilvers and Longhurst 2012), and in other parts of the system, that could enable or thwart the influence of our deliberations on policy processes?

Second, to what extent can success in influencing a particular policy moment like the passage of Edmonton's Energy Transition Strategy be equated with more sustained action by government or other parties? The jury, it must be said, is still out on how much Edmonton's Energy Transition Strategy will succeed in mitigating GHG emissions or spurring energy transition. The strategy, with its focus on market transformation, seeks to initiate and align action across government, markets, and civil society; while important work on this is clearly taking place on the part of City of Edmonton, it is too soon to pronounce on the degree of success it will have, and the key factors influencing this success.

And third, how might policy change and even sustained action by a municipal government along particular lines foster or obstruct wider or deeper changes needed to adequately confront climate change as a systemic issue? Does it push the right levers in the right direction? And what boundaries would we draw in characterizing the broader systems relevant to this question (e.g., political, cultural, economic, biophysical)?

In hindsight, ABCD did not do enough to understand whole systems, systemic change, or social change; we didn't develop the foundations in our research or collaboration to offer robust answers to these three questions. There were some important moments of reflection on social change during the life of ABCD, including a session within an ABCD team workshop where John Gaventa introduced the Power Cube, a tool for analyzing forms, spaces, and levels of power within a system (Gaventa 2006). But we did not, in my view and with the benefit of hindsight, dig deeply enough into accounts of social change, or wrestle with how different stories alive in our group might fit together into a coherent whole. Why weren't we more systemic in our reflection and work?

> *Pressures of time.* One part of this had to do with time scarcity, given the urgency of developing deliberative partnerships, as well as pressures to do extensive research alongside the deliberations (given that our main funding was from a Canadian Social Sciences and Humanities Research Council grant). These pressures tended to fill our workshops and meetings, crowding out bigger picture thinking. Moreover, even if we had found ways to reflect on the big systems

surrounding climate responses in Alberta as the project got under way, we also would have had to dig into systems relevant to each deliberation project.

Lack of ready-to-hand languages and tools. I have suggested, based on my own observations, that engagement in deliberative democracy communities of practice about social change is quite thin. So another challenge in ABCD was a lack of shared language, conceptual tools, and practical tools for mapping power, surfacing and articulating theories of social change, and thinking systemically. This was exacerbated by our annual workshops being co-designed and co-led by shifting teams of practitioners, working pro bono and often stretched for time.

Strength of underlying assumptions about deliberative democracy. In the deliberative democracy community there is a strong current of belief in the value of these processes, and an often fervent energy around seeding more of them as a route to positive social change. As discussed earlier, practitioners and researchers tend to tell stories about deliberative democracy's impacts as being reformist, aggregative, liberal, and very positive (see also C. Lee 2015). Likewise, I suspect that conviction and energy around the value of deliberation on the part of many in ABCD helped keep us from digging deeply and insistently into our implicit accounts of social change and impact. For practitioners and academics alike, witnessing and participating in well-designed citizen deliberation processes build a warranted regard for the intelligence of citizen voices and the power of deliberation to uncover common ground and pathways to action. Yet, though warranted, this appreciation of specific experiences of deliberation and their perceived impacts can impinge on reflection about how deliberative exercises fit into whole systems, and the conditions under which they can support desired systems changes.

The Field of Deliberative Democracy

Some deliberative democracy researchers and practitioners do wrestle with questions around whole systems, systems change, and social change (Atlee 2012; Weymouth and Hartz-Karp 2015). The field as a whole, though, has not made systems a core question, including at field gatherings like those enumerated earlier.

One reason why practitioner gatherings tend not to focus in a sustained way on structural dynamics, system dynamics, or stories of transformational social change is that field-convening organizations want to keep their events—and the field as a whole—open and palatable to conservatives as well as progressives, and talking about social change or systems change or transformative change comes across as progressive (C. W. Lee 2015). A desire to welcome practitioners with diverse ideologies may diminish willingness to look at the widely divergent understandings of social change and dominant systems latent in the field; the temptation is simply to celebrate everyone's good work.

There is also sometimes a perceived conflict between surfacing practitioners' own political convictions or transformational ambitions in their work and a commitment to neutrality or objectivity in their professional roles as facilitators and process designers.[3] In avoiding these issues, the deliberation field risks political inertness, insofar as there are logics and flows in systems that may thwart the impacts of deliberative processes, lead to perverse and unintended outcomes, and lead us to work within limited or unhelpful or unjust frames and boundaries. The articles collected in Lee, McQuarrie, and Walker (2015) outline some of these perversities in the context of public deliberation and economic inequality; earlier, I noted that other perversities can attend focusing engagement on local environmental questions without taking into account the broader systems of which issues being considered may be a mere symptom.

When there is a focus at professional gatherings on the impact and transformative ambitions of the field, attention tends to be on objects of easy agreement: the importance of healing relationships, changing the ways we talk to each other, getting better institutional supports, propagating more dialogue, and diminishing incivility. Yet this can neglect how even these may be symptoms of deeper systemic dynamics.

Learning across Fields

If deliberative democrats are to situate their stories of social change within understandings of whole systems, they can usefully reach for tools from other areas of endeavour and engage theorists and practitioners from other fields. I will briefly outline three such fields and explore the significance of their practices to projects like ABCD and to the deliberative democracy field in general.

Systems Thinking

Systems thinking is a vast area of inquiry and activity. Its proponents urge us to map systems methodically, with an emphasis on building artifacts that externalize mental models and provide common points of reference in dialogue. Such artifacts are used to diagram and model the dependencies, flows, feedback loops, and temporalities of systems. They include causal loop diagrams that map system behaviour by drawing nodes connected by balancing or reinforcing loops (Stroh 2015), and GIGA-maps that trace relationships across many layers and scales, challenging understandings of the boundaries of a problem and the relevant relationships (Systems Oriented Design 2015). Representing systems in diagrams, pictures, and physical models is taken to encourage holistic thinking, as compared to linear prose or purely verbal description and deliberation (Checkland 2000, S22).

A crucial question in depicting a system is deciding its proper boundaries: defining boundaries is contestable, and always linked to a particular purpose. Some systems approaches make a great deal of the need to critically engage with boundary choices, notice underlying sources of selectivity and options forgone, and invite deliberative and collaborative challenges to boundary claims (Ulrich 2005).

My analysis of ABCD, and of deliberative democracy as a more general field, pointed to the need to situate its accounts of the impact of public deliberation in terms of broader systems; the analytical processes just described and their theoretical underpinnings could, I believe, support this, not only in particular projects but in structuring engagement between researchers and practitioners at professional gatherings.

Human-Centred Design

Human-centred design is a second field of inquiry with theories and methods useful to coming to terms with the potential influence of deliberative public engagement on whole systems. Like deliberative democracy itself, human-centred design is a big tent; I will describe it through the work of one of its most prominent exponents, the design firm IDEO (www.ideo.com). Their methodology emphasizes:

Hearing. A team uses qualitative research methods to develop an empathetic understanding of the needs, desires, and aspirations of those for whom they are designing, using interviews and other ways of experiencing the worlds of diverse stakeholders. "At the early stages of the process, research is generative—used to inspire imagination and inform intuition about new opportunities and ideas. In later phases, these methods can be evaluative—used to learn quickly about people's response to ideas and proposed solutions" (IDEO 2011, 32). Here, the goal is to develop a deep understanding of divergence and plurality.

Creating. The team synthesizes and interprets what it has heard and converges on a strategic direction, then again seeks divergence, brainstorming many potential responses to the challenge that has come into focus. The team prototypes some responses through participatory co-design, "building to think, acknowledging that the process of making ideas real and tangible helps us to refine and iterate the ideas very quickly" (IDEO 2011, 83). Prototypes can be models, storyboards, role-plays, or diagrams; they are meant to be quick, cheap, and disposable, designed to validate, communicate, and test ideas. Sharing prototypes within the team and with outsiders supports learning.

Delivering. Based on this learning the team converges on mini-pilots that are taken out into the world, still with low investment and with a readiness to learn through failure. Ongoing evaluation and measurement supports learning, and pilots are repeated until the team has feasible, sustainable interventions that respond to stakeholder needs.

Human-centred design emerged from fields like architecture, user interface design, and industrial design; more recently, it is associated with social innovation and an orientation to whole systems (Jones 2014). Applied within deliberative democracy projects and communities, it could more systematically bring new voices and perspectives into design and reflection; externalize understandings of systems to build understanding and alignment; and enable low-investment experiments to advance understandings of how to intervene successfully in systems.

Systemic Design

Systemic design combines the two approaches just described. The mindset, methodology, and methods of systemic design aim to address wicked problems: to enable "diverse teams to develop an elevated perspective of the challenge and translate novel insights into rapid action" while accelerating learning (Ryan 2014, 12). Ryan (2014, 6) describes the mindset of systemic design as "inquiring, open, integrative, collaborative, and centred." The methodology or abstract logic of systemic design involves:

Inquiring. Moving beyond the knowledge held by the group by using stakeholder ethnography, literature reviews, engagement with experts, and learning journeys that take the group to parts of the system they may not have experienced before.

Framing. Bringing into view how issues and solutions are implicitly being framed or bounded, considering alternative frames, and developing a shared frame.

Formulating. Having the group articulate the diverse values that are motivating their work on an issue, and creating tangible artifacts (diagrams, extensive maps, physical models) that support a common understanding of values, frames, and understandings of the system.

Generating. Taking these artifacts out into the world to see how others respond, and perhaps as actual prototypes of interventions in the system. These artifacts should be quickly and cheaply produced, so that multiple understandings and interventions can be tested and build learning, including through failure.

Facilitating. Establishing and supporting norms for working together, and planning and creating settings and dialogues where the group can invite others into the work.

Reflecting. Assessing the effects of the group's actions in the world, and moving from diverse observations about these to shared understandings that support further cycles of analysis and action.

The methods of systemic design draw from both systems thinking and design thinking: they include creating rich pictures of systems, mapping systems, and diagramming causal loops, and practices of prototyping that enable interventions in systems to be created quickly and roughly and tested on the ground (Ryan 2014).

How Might these Three Fields Inform and Complement Theories and Practices of Deliberative Democracy?

In thinking about the evolution of ABCD as a project, several elements of systems theory, human-centred design, and systemic design stand out as potentially helpful. First is the primacy in these approaches of systemic understanding as a foundation for effective action. In its early stages, ABCD used deliberative methods, both within our emerging team and with larger groups of stakeholders, to consider how citizen involvement could advance climate responses in Alberta. However, we did not methodically map broader systems relevant to our work—for example, the political, social, cultural, and economic systems that produced provincial and municipal climate policy in Alberta, with their complex dynamics, feedback loops, and webs of interconnection. And we did not engage in methodical "boundary critique" in our assessment of potential interventions. Views of these issues were always in play but often implicitly, inchoately, and without alignment of understanding or purpose across the ABCD team.

The conceptualization of complex, interdependent systems offered earlier in this chapter and the methods of system mapping just outlined could have been an important foundation for our collective work. This work could have been done iteratively in the cross-sectoral workshops that developed ABCD as a project, the team meetings of ABCD, and in meetings of smaller subgroups and teams. Incorporating systems analysis into the development of ABCD would have helped us to understand the potential impact of mini-publics on the climate policy system in Alberta; might have led us to articulate mini-publics differently with social movements and political opportunities (Kahane and MacKinnon 2015, 18–20; Kenrick 2013); or might have steered us to strategies for systemic influence other than mini-publics. Systems analysis would have mapped the forces, players, causalities, feedback loops, and temporalities within which we were intervening; it would have helped us to align around a shared, comprehensive analysis, and to critically assess the strategies we should pursue.

A second element of these three fields that stands out as helpful is how they deliberately move back and forth between divergence and convergence. Deliberative democratic methods are sophisticated in assembling diverse groups, surfacing values, weighing trade-offs, and using group processes to converge on common ground. In systems thinking, human-centred design, and systemic design, however, the movement between divergence and convergence is more crisply captured in methodology, and there are multiple stages of divergence-convergence; this methodology would have been useful to us in developing ABCD, and in discouraging what feel in retrospect like moments of premature closure.

Third, the systemic methods discussed offer a suite of ethnographic approaches to understanding diversity within systems. In addition to bringing diverse groups into deliberations (which is what ABCD mainly did), human-centred design and systemic design emphasize interviewing, learning journeys, and other methods of hearing and inquiring. ABCD, which was a very white and in other ways relatively homogeneous team, would have benefited by adding some of these approaches to our work (see chapter 2 for the importance of social diversity to climate deliberation).

Fourth, I believe that ABCD would have benefited from creating artifacts and prototyping. As brought out in chapter 6 of this volume, ABCD struggled both within its membership and in partnership development to communicate the distinctiveness of deliberation as a mode of public engagement; "building to think" would have been powerful in both articulating our methods to ourselves and communicating them to others. It not only would have clarified and aligned our thinking within ABCD but would have given us pictures, diagrams, and other artifacts to communicate our thinking to others, and to support others in situating themselves in relation to our approach. One tool we did use repeatedly in ABCD to create artifacts of our thinking was graphic recording—depicting conversations on large sheets of paper during deliberative events (see chapter 1): for all of the virtues of graphic recording, though, this method is importantly different from those offered by design approaches. With graphic recording, artifacts are created by a professional rather than participants; there is one artifact rather than many for a given segment of work; and the artifacts are professional-looking and permanent rather than "quick-and-dirty" and easily revised by participants (for an extended discussion of graphic recording see C. Lee 2015, 123–49).

And fifth, the deliberation projects that ABCD brought into the world tended to be high stakes: our three citizens' panels took months or years to develop, involved intense labour by teams, and were expensive. This stands in contrast to an emphasis on low-stakes, quick, iterative learning by doing. It's not that this "mini-pilot" approach could transfer straightforwardly to all of the contexts in which ABCD worked—it might not, for example, have fit the needs of the City of Edmonton that gave rise to that Edmonton Panel. But it is interesting to think about points in ABCD's formation as a project when diverse, low-stakes deliberative interventions might have supported us in clarifying our thinking, learning about systems, and converging on strategy. We could have designed small, rough-around-the-edges deliberations with clear learning purposes in relation to our bigger project, and folded this learning back into our methodologies and our development of partnerships.

None of these insights from other fields is a silver bullet; rather, they hint at new possibilities in deliberative democratic practice. And it is important to recognize obstacles to their use. When I canvassed reasons why ABCD did not delve deeply and persistently into questions of social and systems change, I mentioned the lack of ready-to-hand language and tools: these other fields have value to offer here in the methodologies and methods I've outlined. I also mentioned the strength of underlying assumptions about deliberative democracy on the part of some members of ABCD; these other approaches to systems change might usefully have helped us to articulate these assumptions, assess them critically, and bring them into explicit relationships to the particular systems in Alberta that we were seeking to affect. The third reason I cited, though, had to do with scarce time on the part of a mainly volunteer ABCD team, pressures associated with the exigencies of particular projects, and constraints that arose from complexities of partnerships and policy processes. My enthusiasm about bringing deliberative democracy together with systems theory, human-centred design, and systemic design is tempered by an awareness of the crush of such circumstances in projects like ours.

Stepping back from the particularities of ABCD, I believe that the mindsets, methodologies, and methods of systems theory, human-centred design, and systemic design can contribute importantly to deliberative democracy as a field. These problem-solving approaches offer a range of tools that could be used in citizen deliberations, as well as in processes of project development: in both contexts, they would support more careful embeddedness of processes in systems, and more careful analysis of potential impacts, than current deliberative democracy approaches typically achieve. These problem-solving approaches also would be useful in building field learning, since these mindsets, methodologies, and methods might support deliberative democracy researchers and practitioners in thinking concertedly about questions of impact, social change, and systems change.

What conditions would be needed for these new kinds of reflection to enter deliberative democratic theories and practices? In some ways, the conditions exist: the tools could be brought into sessions at professional gatherings, including by invited practitioners of systems approaches; they could structure workshops focused on field learning; and if ready-to-hand tools were developed, they might be taken up in particular projects. Three dynamics that may cut against deliberative democrats picking up these tools are the time and resource pressures of the work, challenges to assumptions about the virtue and effectiveness of the work that some might find uncomfortable, and the professionalization of the field, which may incline practitioners to tout the effectiveness of their tools rather than engaging critically about whether the tools are in fact effective in shifting systems (Kahane and Loptson 2017).

How Might Deliberative Democracy Inform and Complement Systems Theory, Human-Centred Design, and Systemic Design?

The cross-fertilization of deliberative democracy with systems theory, human-centred design, and systemic design has potential in the other direction as well. I have been struck, in my forays into these fields, by the limited exploration of questions of democratic involvement and accountability. To the extent that these approaches engage with democratic publics and citizens, it tends to be in the language of clients, customers, or stakeholders. These terms resonate uncomfortably with neoliberal understandings of citizens as "users and choosers" rather than "makers and shapers" of social and political policy (Cornwall and Gaventa 2001). Insofar as there is an elitist tinge to the three approaches to systems change that I've outlined—a deep ethnographic sensibility, but on the part of a privileged cohort of ethnographers—deliberative democracy can offer both analysis grounded in viewing citizens as key agents in policy development and political change, and practical tools for rooting these approaches more firmly in the will and activity of democratic publics.

Furthermore, deliberative democrats have thought and practised intensively around the pragmatic as well as principled connections between work with citizens and influence on governments. While I suggested above that the eight deliberative democratic change stories should be nuanced through a systems lens, it's also true that systems theory, human-centred design, and systemic design might benefit from wisdom embedded in these eight stories when it comes to securing influence for social change processes.

Finally, the strongly normative tendencies of deliberative democratic theory may be a useful counterpoint to the scientific and commercial roots of systems theory and user-centred design, reminding us of the ethical implications of particular methods and offering a rich conceptual and analytical language for thinking about ethics in the context of democratic intervention in complex systems.

Conclusion

ABCD partnered with organizations in the province to convene citizens: our goal was to enhance climate responses through public participation and to advance learning about deliberative democracy and climate change. I have suggested that ABCD, for all its strengths, would have benefited from organizing its work more methodically and consistently around questions of systems change, in terms of the orientation of the overall project, the development of particular partnerships, and the design of citizen deliberations. This in itself has been a key part of our learning.

Climate change is not only a wicked or super-wicked problem, it is a systemic one. The more deliberative democracy can foreground questions of systems change in mindsets, methodologies, and methods, the more helpful it can be in building effective political, social, and cultural responses to climate change and other systemic questions. A systemic deliberative democracy would support citizens in coming to grips with the wicked and socially complex character of climate change, so that they could shape wise and effective responses to the challenges. Such an approach would support governments and other bodies seeking to convene citizens around climate change in themselves coming to see the challenge through a systems lens.

As deliberative democrats become more adept at working with systemic mindsets, methodologies, and methods they will position themselves to infuse a greater citizen voice in quarters where a systems perspective is already present but where greater democratic engagement is needed. A systemic deliberative democracy would support efforts not just to interpret the whole system, but to change it.

Notes

1. Alex Ryan uses "systemics" to describe an ensemble of the fields of systems thinking, human-centred design, and systemic design.

2. Evidence of the Edmonton panelists' greater confidence could be found in the participant's response to the survey question: "How much can people like you affect what the government does?" Participants were given a five-point scale on which to place their answer where 1 was "not at all," 2 was "a little," 3 was "a modest amount," 4 was "a lot," and 5 was "a great deal." The average answer in the pretest was 2.89 whereas the average taken after session 6 was 3.38 (Boulianne and Loptson 2013).

3. The question of practitioner neutrality and political commitment was taken up (with difficulty but also success) in the 1980s and 1990s in the field of dispute resolution; I believe that there are lessons for deliberative democrats in that experience. See Bailey 1991; Bryan 1992; Lederach 1995; and Merry 1987.

References

Alberta Climate Dialogue and Centre for Public Involvement. 2015. "Impacts of the Citizens' Panel on Edmonton's Energy and Climate Challenges." Alberta Climate Dialogue Working Paper. http://www.albertaclimatedialogue.ca/wp-content/uploads/2015/03/Impacts-of-Edmonton-Climate-Panel.pdf.

Andrais, Jim. 2015. Telephone Interview with Lorelei Hanson, December 9.

Armson, Rosalind. 2011. *Growing Wings on the Way: Systems Thinking for Messy Situations*. Axminster, UK: Triarchy Press.

Atlee, Tom. 2012. *Empowering Public Wisdom: A Practical Vision of Citizen-Led Politics*. New York: Random House.

Bailey, F.G. 1991. "*Tertius Lucans*: Idiocosm, Caricature, and Mask." In *Conflict Resolution: Cross-Cultural Perspectives*, edited by Kevin Avruch, Peter W. Black, and Joseph A. Scimecca, 63–89. New York: Greenwood Press.

Bohman, James. 1996. *Public Deliberation*. Cambridge: MIT Press.

Bryan, Penelope E. 1992. "Killing Us Softly: Divorce Mediation and the Politics of Power." *Buffalo Law Review* 40: 441–523.

Checkland, Peter B. 2000. "Soft Systems Methodology: A Thirty Year Retrospective." *Systems Research and Behavioral Science* 17: S11–S58.

Chilvers, Jason, and Noel Longhurst. 2012. *Participation, Politics, and Actor Dynamics in Low Carbon Energy Transitions*. Norwich, UK: University of East Anglia Science Society and Sustainability.

Cornwall, Andrea, and John Gaventa. 2001. "From Users and Choosers to Makers and Shapers: Repositioning Participation in Social Policy." *IDS Working Paper* 127. Brighton, UK: Institute of Development Studies.

Cutler, Fred, Richard Johnston, R. Kenneth Carty, Andre Blais, and Patrick Fournier. 2007. "Deliberation, Information and Trust: The British Columbia Citizens' Assembly as Agenda Setter." In *Designing Deliberative Democracy: The British Columbia Citizens' Assembly*, edited by Mark E. Warren and Hilary Pearse, 166–91. Cambridge: Cambridge University Press.

Delli Carpini, Michael X., Fay Lomax Cook, and Lawrence R. Jacobs. 2004. "Public Deliberation, Discursive Participation, and Citizen Engagement: A Review of the Empirical Literature." *Annual Review of Political Science* 7: 315–44.

Easterbrook, Steve. 2011. "The Power to Change Systems." October 16. http://www.easterbrook.ca/steve/2011/10/the-power-to-change-systems/.

Fagotto, Elena, and Archon Fung. 2009. "Sustaining Public Engagement: Embedded Deliberation in Local Communities." Occasional Research Paper, Everyday Democracy and the Kettering Foundation. http://www.everyday-democracy.org/sites/default/files/attachments/Sustaining-Public-Engagement.pdf.

Fung, Archon. 2003. "Survey Article: Recipes for Public Spheres: Eight Institutional Design Choices and Their Consequences." *Journal of Political Philosophy* 11(3): 338–67.

Gastil, John, and Peter Levine. 2005. *The Deliberative Democracy Handbook: Strategies for Effective Civic Engagement in the 21st Century*. San Francisco: Jossey-Bass.

Gaventa, John. 2006. "Finding the Spaces for Change: A Power Analysis." *IDS Bulletin* 37: 6.

Gaventa, John, and Gregory Barrett. 2010. "So What Difference Does It Make? Mapping the Outcomes of Citizen Engagement." *IDS Working Paper* 347. http://www.ids.ac.uk/files/dmfile/Wp347.pdf.

Gutmann, Amy, and Dennis Thompson. 1996. *Democracy and Disagreement.* Cambridge: Harvard University Press.

Hannah, Scott, and Gerard MacLellan. 2015. Personal Interview with Author, April 29.

Henderson, Ben. 2015. Interview with author.

Hobbs, Lyndsay. 2013. "Participant Experiences of Change in Deliberative Setting." ABCD Working Paper Series. *Alberta Climate Dialogue*. October 2013. Accessed November 16, 2015. http://www.albertaclimatedialogue.ca/participant-experiencs-of-change-in-a-deliberative-setting/participant-experiences-of-change-in-a-deliberative-setting/.

IDEO. 2011. "Human Centered Design Toolkit." *IDEO*. Accessed October 27, 2015. https://www.ideo.com/by-ideo/human-centered-design-toolkit.

Iveson, Don. 2015. Edmonton City Council Meeting, April 29.

Johnson, Genevieve Fuji. 2015. *Democratic Illusion: Deliberative Democracy in Canadian Public Policy*. Toronto: University of Toronto Press.

Jones, Peter. 2014. "Systemic Design Principles for Complex Social Systems." *Translational Systems Sciences* 1: 91–128.

Kahane, David, and Kristjana Loptson. 2017. "Academics as Deliberation Practitioners." In *The Professionalization of Public Participation*, edited by Laurence Bherer, Mario Gauthier, and Louis Simard. London: Routledge.

Kahane, David, and Mary Pat MacKinnon. 2015. "Public Participation, Deliberative Democracy, and Climate Governance: Learning from the Citizens' Panel on Edmonton's Energy and Climate Challenges." *CISDL/GEM Working Paper Series on Public Participation and Climate Governance.* http://cisdl.org/public/docs/KAHANE.pdf.

Kenrick, Justin. 2013. "Emerging From the Shadow of Climate Change Denial." *ACME: An International E-Journal for Critical Geographies* 12(1): 102–30. http://ojs.unbc.ca/index.php/acme/article/view/955/809.

Klein, Naomi. 2014. *This Changes Everything: Capitalism vs. The Climate*. New York: Simon and Schuster.

Lederach, John Paul. 1995. *Preparing for Peace: Conflict Transformation Across Cultures*. Syracuse: Syracuse University Press.

Lee, Caroline W. 2015. *Do-It-Yourself Democracy: The Rise of the Public Engagement Industry*. Oxford: Oxford University Press.

Lee, Caroline W., Michael McQuarrie, and Edward T. Walker. 2015. *Democratizing Inequalities: Dilemmas of the New Public Participation*. New York: NYU Press.

Lee, Marc. 2015. "A Conversation on Climate Justice." *Policy Note: A Progressive Take on BC Issues*. http://www.policynote.ca/a-conversation-on-climate-justice/.

Levin, Kelly, Benjamin Cashore, Steven Bernstein, and Graeme Auld. 2012. "Overcoming the Tragedy of Super Wicked Problems: Constraining Our Future Selves to Ameliorate Global Climate Change." *Policy Sciences* 45: 123–52.

Lukensmeyer, Carolyn J. 2012. *Bringing Citizen Voices to the Table: A Guide for Public Managers*. San Francisco: Jossey-Bass.

Mansbridge, Jane, James Bohman, Simone Chambers, Thomas Christiano, Archon Fung, John Parkinson, Dennis F. Thompson, and Mark E. Warren. 2012. "A Systemic Approach to Deliberative Democracy." In *Deliberative Systems: Deliberative Democracy at the Large Scale*, edited by John Parkinson and Jane Mansbridge, 1–26. Cambridge: Cambridge University Press.

Merry, Sally. 1987. "Disputing Without Culture." *Harvard Law Review* 100: 2057–73.

Mutz, Diana C. 2008. "Is Deliberative Democracy a Falsifiable Theory?" *Annual Review of Political Science* 11(1): 521–38.

Nabatchi, Tina, and Matt Leighninger. 2015. *Public Participation for 21st Century Democracy*. San Francisco: Jossey-Bass.

Pembina Institute and HB Lanarc. 2012. Edmonton's Energy Transition: Discussion Paper. http://www.edmonton.ca/city_government/documents/PDF/Edmonton_Energy_Transition_Discussion_Paper.pdf.

Rittel, Horst, and Melvin Webber. 1973. "Dilemmas in a General Theory of Planning." *Policy Sciences* 4: 155–69.

Row, Jesse. 2015. Telephone Interview with Lorelei Hanson, Deborah Schrader, Mary Pat McKinnon and David Kahane, June 7.

Ryan, Alex. 2014. "A Framework for Systemic Design." *FORMakademisk* 7(4). https://journals.hioa.no/index.php/formakademisk/article/view/787.

Ryfe, David M. 2005. "Does Deliberative Democracy Work?" *Annual Review of Political Science* 8(1): 49–71.

Sterman, John. 2000. *Business Dynamics: Systems Thinking and Modeling for a Complex World*. Boston: McGraw-Hill Education.

Stroh, David Peter. 2015. *Systems Thinking for Social Change*. White River Junction, VT: Chelsea Green.

Systems Oriented Design. 2015. "GIGA-Mapping." *SOD*. http://www.systemsorienteddesign.net/index.php/giga-mapping/giga-mapping-information.

Ulrich, Werner. 2005. "A Mini-Primer of Boundary Critique." http://wulrich.com/boundary_critique.html.

van der Lans, Dymphna. 2014. "How Systems Thinking Can Impact Climate Change." *Clinton Foundation*. September 19. https://www.clintonfoundation.org/blog/2014/09/19/how-systems-thinking-can-impact-climate-change.

von Lieres, Bettina, and David Kahane. 2007. "Inclusion and Representation in Democratic Deliberations: Lessons from Canada's Romanow Commission." In *Spaces for Change: The Politics of Participation in New Democratic Arenas*, edited by Andrea Cornwall and Vera Schattan Coelho, 131–51. London: Zed Books.

Weymouth, Robert, and Janette Hartz-Karp. 2015. "Deliberative Collaborative Governance as a Democratic Reform to Resolve Wicked Problems and Improve Trust." *Journal of Economic and Social Policy* 17(1): art. 4. http://epubs.scu.edu.au/cgi/viewcontent.cgi?article=1325&context=jesp.

Williams, Melissa. 1998. *Voice, Trust, and Memory: Marginalized Groups and the Failings of Liberal Representation*. Princeton: Princeton University Press.

Young, Iris Marion. 2005. *Inclusion and Democracy*. Oxford: Oxford University Press.

Conclusion

The Potential of Deliberation to Tap the Power of Citizens to Address Climate Change and Other Issues of Sustainability

Tom Prugh and Matt Leighninger

There are signs that twenty-first century public institutions are not up to the challenge of dealing with wicked problems like climate change. For this failing, and a host of other reasons, the trust and confidence citizens once had in their public institutions is in sharp decline. If citizens no longer believe that the democratic structures and processes currently in place are capable of addressing one of the most pressing problems of our time, an opportunity to adopt new tools and methods is present.

This book is a product of the strong research component built into the Alberta Climate Dialogue (ABCD) project. It assembles a rich compilation of theoretical insight and practical wisdom from nine contributors with expertise in deliberative practice and sustainability issues such as climate change, as well as close familiarity with Alberta's communities. The contributors' chapters offer a great deal of nuanced analysis and reflection, and although we cannot hope to capture all of it, in this concluding chapter we aim to extract some of the key themes and observations toward making sense of a complex whole. Then, drawing on the experiences with the ABCD exercises addressing climate change among Albertans, we briefly explore the role deliberation might play in confronting the host of sustainability problems facing not only the citizens of Alberta but all of humanity, and argue that deliberation should find a natural home in the increasingly activist urban- and community-centred sustainability movement. We close with a tempered call for "amateurism," in the traditional sense of work by engaged and knowledgeable non-professionals, in deliberation.

Key Themes

Here are some themes and ideas that emerge from the previous chapters, with an emphasis on those that might particularly interest practitioners and concerned citizens. Where applicable, references to chapters in parentheses indicate where more material on a particular theme can be found.

> *Deliberation is not just for experts.* Deliberation needs to involve—and be useful to—a wide range of people with different values, concerns, life stories, and world views. This is especially true when it is used to address problems such as climate change, which are complex and affect essentially everyone. Deliberation can serve to integrate those differing perspectives and values, and thus support citizens in expanding their circle of concern as well as, crucially, stimulating and organizing input on the condition of their community and the ecological systems that enable its existence (introduction and chapter 2). As the product of a research effort, this book may appear to frame deliberation as an arcane and delicate practice, organized by experts, in which ordinary people can only participate if they are given ample preparation. Indeed, the experts have crucial roles to play, one of which is to ameliorate the tension between the complexities explored in these pages and the need to bring deliberation down to earth and engage a much wider public. But one of the more remarkable aspects of the ABCD experiences, and an enormous group of other deliberation stories, is that ordinary people can accept and adopt the practices of organized deliberation when they are properly introduced to them, despite the lack of such activities in most day-to-day political environments. We will have more to say about "deliberation for the people" in the last section below.

> *Deliberation works best in an oxygen-rich atmosphere.* That is, not in a vacuum; it should support action and be tied to policy outcomes. People take to it with surprising enthusiasm, but it is valuable to ensure that the process leads to action. Deliberation exercises may be undertaken in the absence of such links, as indeed the history

of democratic deliberation repeatedly reveals. But that is a waste of civic capital. Deliberation events and processes ideally should not be used merely to generate support for a predetermined policy, or even to select from a menu of options. They should instead be designed so that people can provide meaningful input into the range of potential policy options, and so they can decide how to contribute their own time and energy to implementing solutions. One of the strengths of democratic deliberation is that it taps the knowledge and values of a body of people with a stake in the outcomes but who are not often consulted—except to the extent that their votes are sought and their favour curried at election time. (See chapter 3 for a discussion of the ABCD experience in Alberta.)

Framing is more than decoration. How issues are framed and presented to participants can alter, for good or ill, the conclusions they reach. Framing for deliberation should present and clarify the different ways of looking at an issue so people can compare them fairly in order to weigh appropriate courses of action (see chapter 5).

Climate change, for instance, is usually framed as a challenge to be mitigated with technological solutions (the so-called "ecomodernist" stance), whereas in fact it may require deeper social and behavioural change. That is, while it is typically presented as a problem to be solved—a big, complex one to be sure—in fact it may by now have become largely a predicament that can only be coped with by means of various adaptations. (We might term this the "ecotransitionist" frame.) Adopting this latter frame immediately raises major, serious questions about social justice; topping the list might be how to help people and nations that bear little or no responsibility for climate change yet are suffering disproportionately from it.

At the same time, adaptation as a frame and strategy situates the problems of addressing climate change in particular places, which is an argument for localism and types of governance well suited to communities—such as deliberation. (In addition to these two frames, there is at least one more, which might be labelled "extreme adaptation." This frame is based on the growing sense among some observers that radical resource scarcity will demand a deep retrenchment in our everyday technologies and a reversion to simpler lifestyles. We discuss this idea further below.) Organizers and participants in deliberation exercises need to be alert to the frames participants bring to the table, and also to their effects on shaping the process and the suite of policy options considered.

No deliberation without representation. Well, not *no* deliberation— but any deliberation structure, whether a one-off event or a standing body, needs to give due attention to the issue of representativeness: how closely the mix of participants resembles the larger community from which they come (see chapter 4). There are a number of ways of doing this, from an exact polling-style approach to a more welcoming, inclusive strategy that tries to achieve a turnout that is both large and diverse. At one end of the scale, random-sample methods try to create a more or less perfect microcosm of the community; at the other end, organizers welcome all comers but spend a disproportionate amount of time trying to reach people who seem less likely to participate. Either way, deliberative processes usually have to involve or at least influence large numbers of people in order to have an impact on policy. This is where random-sample "mini-publics" often fall short, since they don't produce the critical mass of participants or the political will necessary for action. However, they can be valuable components of a broader strategy. As the ABCD projects illustrated, achieving adequate representation can be hard to do well, for a variety of reasons. Self-selection of participants is a factor even in random-sample strategies, and most deliberation exercises tend to over-represent people with higher education while under-representing young people. But to the extent that representativeness can be achieved, it enhances legitimacy and maximizes the odds of introducing into the deliberation the richest range of values, problem perspectives, and possible solutions.

Trust but verify. Deliberation exercises frequently (invariably?) become crucibles in which different actors with widely varying aims, expectations, and interests come together. While deliberation can be a powerful means of supporting collaboration, it's most likely to succeed if trust and respect for different contexts and cultures of risk is built carefully (see chapter 6). This takes time. Deliberation is ultimately about power: who exercises it, and to what ends. When successful, deliberation leads to policy decisions, to inputs that shape policy, or to volunteer-driven action efforts in which people work together to implement the ideas they have generated (or all three). No matter what kind of outcome you want to support, the stakes can be significant and the process delicate. Trust building is crucial to keeping that process civil and productive.

It's complicated, but it's simple. Considered in the full richness of the associated scholarship and practice traditions, deliberation can be a complex business, from what it signifies and embodies in terms of political theory to the nuances of recruitment, process design, and competent facilitation. This makes it hard work (see chapter 7). Likewise, climate change and other sustainability issues are complex, global problems, frequently termed "wicked": different stakeholders define the issues in different terms; understanding of the problems changes over time; there may be no clear "right" solutions; and what appears to be a problem may be just a symptom of something deeper. Yet climate change manifests itself in characteristically local effects, and deliberation itself is also "particular and local" (see chapter 8): it takes place in a specific community and usually focuses on a narrowly defined issue. This creates an opportunity. Deliberation asks—and enables—citizens to confront complexity (in any issue, not just climate change) and, if not master it, at least become acquainted with it; to grapple with issues, to sit with them and become conversant with their nuances; and to make thoughtful and reasoned judgments about how a community ought to address them. (See "Deliberating Cities and Communities," below)

Shelter Needed from the Perfect Storm

The ABCD deliberation exercises offer hopeful evidence that deliberation can be a useful, perhaps necessary, method for confronting multiple, complex, and even existential challenges. Climate change certainly qualifies as one of those: it is deranging the most complex system of which we know—the Earth's biosphere—thereby threatening the viability of civilization in ways that we barely understand and with emergent consequences we cannot predict. The litany of likely (and indeed already observed) effects of a warmer world is by now familiar: rising sea levels; hotter and longer droughts; heavier floods; wilder weather and more extreme storms; stressed and unreliable fresh water supplies;

ecosystems corrupted by invasive species or destroyed altogether; expansion of disease vectors; degradation and possible collapse of marine food chains as the oceans acidify; loss of agricultural productivity; and so on.

These problems alone would make the governance challenges of the coming decades daunting enough, but they are not the only systemic changes coming at us fast. At least two others are visible on the horizon.

The first is the decline and approaching end of the fossil fuel era. At this writing, gasoline prices in North America are low and sales of SUVs are surging; arguments about "peak oil" are laughed off or ignored. Nevertheless, while short-term fluctuations in energy prices and the vagaries of geopolitics may temporarily mask the longer trends, the fact remains that humanity for the last 250 years or so has been burning through an endowment of fossil energy created over eons by geological forces; such a windfall will not come again. The early signs of trouble include the increasing expense and difficulty of finding oil deposits to replace current consumption. Rising demand and the exhaustion of conventional oil supplies force oil companies to develop sources such as Alberta's tar sands, fields in the Arctic Ocean, and those far beneath the deep sea floor. Not only are these deposits more costly, dangerous, environmentally destructive, and challenging to tap, they simply do not yield useful energy products at the rate conventional fields once did. The amount of energy they yield for the energy required to get it out of the ground, refine it, and deliver it to consumers—a critical ratio called EROI (energy return on investment)—has plunged over the last century or so from roughly 100:1 to less than 30:1, and even lower in many cases. That matters, because the energy available to run our cars, planes, trains, and ships is only that which is left over once the energy development bill has been paid.

The EROIs of coal and natural gas have also been declining in recent years. Add to that the growing urgency of leaving fossil fuels in the ground unburnt so as to avoid the serious risk of catastrophic climate change, and the urgent compulsion to end the fossil fuel era becomes plain. However, that is easier said than done. A debate rages among environmental and energy scholars, scientists, and activists about whether and how fast renewable sources of energy can be substituted for fossil energy, but nobody argues that it will be easy. Building out a new energy regime will cost trillions of dollars and take many years, and of course the energy to do so must come from fossil fuels themselves. Moreover, it is an open question whether all sectors can be engineered to function on renewables. Lighting and conditioning buildings would be relatively straightforward using renewably generated electricity, but high-heat industrial processes are not so easily tackled, and there are serious obstacles to transforming global transportation—almost completely dependent on energy-dense liquid fuels—to run on renewables. No current or foreseeable biofuel or renewable electricity source is available in sufficient quantities to drive the trains, ships, commercial aircraft, and heavy trucks that current developed-world economies rely upon. And while nuclear power has its dogged champions (including many ecomodernists), it faces nearly insurmountable obstacles of its own: waste, safety and security issues, huge costs, long lead times, and popular opposition.

Finally, it is well worth noting that no society has ever fully transformed its energy regime. As the Canadian energy analyst Vaclav Smil has amply documented, new energy sources have not eliminated old ones (whale oil possibly excepted) but rather have been added into the mix as humanity's collective energy consumption has soared over the last few centuries (Smil 2010). Yet the challenge of the renewable transition is to *displace* the overwhelmingly primary source of energy—fossil fuels—with something quite different.

The upshot is that, barring cold fusion or some other miracle, the voracious consumption of energy that underpins the current global economic system is probably unsustainable, even apart from its effects on the climate. In the not too distant future we will have to make do with less energy as the one-time pulse of cheap and abundant fossil fuels that supports modern civilization—and hundreds of millions of newly middle-class people—tails off and ends. That seems likely to usher in a period of social and political unrest.

The second systemic challenge to the current order is intertwined with the energy dilemma: a range of developments that suggest the approaching end of economic growth itself. Since energy availability underlies economic growth, diminishing energy supplies will clearly impede growth, but there are other factors at work too. Ecological economists such as Herman Daly and many others have argued for years that infinite economic growth on a finite planet is impossible anyway (see, for instance, Daly 1991). But now economists with more mainstream orientations are also beginning to talk about "headwinds"— declining rates of innovation, demographic factors, globalization, wealth and income inequality, vast government and private debt—in seeking explanations for Japan's long stagnation and the globally weak recovery from the 2008 crash (Galbraith 2014; Gordon 2012).

Growth has long been the go-to solution for many or most political problems, so its decline and end seem likely to add to the stresses on society imposed by declining energy. While it is possible, in terms of the Earth's resource availability, to provide decent lives for most people on the planet—to have "prosperity without growth" (as the title of one prominent study puts it; see Jackson 2009; Victor and Jackson 2015)—this admirable goal will remain far out of reach as long as existing resources and wealth are so unevenly apportioned among the world's peoples. Serious issues of adjustment and wealth distribution remain to be negotiated as the era of growth winds down (Heinberg 2011).

To sum up, the world is changing in ways that challenge our usual assumptions about humanity's economic future and that could require profound shifts in the shape and character of our communities, our economies, and our methods of governance. The end of a stable climate, along with the end of the unique and extraordinary period of cheap and abundant energy and the probable end of economic normalcy, together could spell the end of political normalcy. There is an urgent need to build governance systems that can adjudicate what are likely to be increasingly contentious disputes over how to navigate these challenges.

Deliberating Cities and Communities

Could a culture of democratic deliberation help? It remains to be seen, but the question may have an answer in the near future. To date, democracies' performance in addressing climate change and other sustainability issues has, on the whole, been disappointing (notwithstanding the somewhat toothless agreement struck in Paris in December 2015). We suspect that a key reason lies in an inherent weakness of representative democracies: they isolate their citizens from each other as political actors, and from direct confrontation with the problems governance is meant to solve, by treating them essentially as wards or children (Kemmis 1990; Nabatchi and Leighninger 2015). In this way, modern democracies tend to cultivate what philosopher Richard Weaver calls "a sort of contempt for realities" (cited in Orr 2013, 287). Even when polls reveal widespread support for more aggressive action on climate change, the ordinary machinery of democracies tends to provide few potent means to convert it to action.

But perhaps the spread of deliberative civic engagement (DCE) could help change that. In conducive settings, deliberation changes minds, helps viewpoints evolve, and improves the quality of collective decision making—processes that urgently need to be promoted with respect to sustainability issues. Deliberation is also tailored to local concerns and interests, which "dictates environmental watchfulness and, when problems arise, a deliberate search for solutions," as well as helping to resist private interests whose actions may be inimical to sustainability (Gundersen 1995, 200).

Successful DCE initiatives, which have sprung up around the world—Australia, Brazil, China, India, Nigeria, the Philippines, South Africa, and in Europe and North America—tend to share certain characteristics:

> they bring together a large and diverse group of citizens . . .
> in structured and facilitated small-group discussions combined with larger groups focused on action, plus they create . . .
> the opportunity for participants to consider a range of arguments, information, and policy options, and . . .
> they focus on concrete outcomes. (Leighninger 2012, 20)

Like the exercises members of ABCD were involved in, most of these DCE initiatives have been ad hoc, but there are a number of examples of sustained deliberative engagement as well, both historical and contemporary. Particularly in Brazil and other parts of the Global South, deliberative engagement has been built into the way that many cities operate. These instances of sustained engagement include citizen-driven land use planning exercises in India, local health councils in Brazil, ward committees in South Africa, "co-production" in the Philippines, and annual participatory budgeting processes in hundreds of cities (Spink and Best 2009; Peixoto 2012). In some of these cities, tens of thousands of people are engaged annually.

In addition to giving people meaningful opportunities to take part in public decision making and problem solving, these examples of sustained engagement have been connected with other societal outcomes, such as higher tax compliance, lower levels of corruption, higher trust in government, higher levels of economic development, and lower economic inequality (Touchton and Wampler 2014). These kinds of outcomes may be due to the fact that sustained engagement strengthens social capital and the web of relationships between neighbours.

DCE remains a largely local phenomenon. This is particularly true of sustained forms of engagement. However, the spread of online networks, especially the hyperlocal online networks that have proliferated dramatically at the neighbourhood and town level in recent years, provide new opportunities for scaling up engagement to address global challenges like climate change. On any level, DCE tends to have the greatest impacts when it involves a large, diverse critical mass of participants; the sheer number of participants is what helps give these processes the political weight to affect policy makers inside government and/ or the accumulated volunteer capacity to implement action ideas outside government (Nabatchi and Leighninger 2015).

When it achieves this kind of scale, DCE has much to offer as a way for communities to come to grips with complex problems, such as climate change, that are both universal and particular. Interestingly, the spread of deliberation coincides with an impulse toward the localization of responses to sustainability problems. Climate change mitigation and adaptation are increasingly being adopted into the policy portfolios of cities and local communities worldwide, driven partly by disappointment with the pace of progress at the international level (Worldwatch Institute 2016). Cities of all sizes and on every continent are committing publicly and in writing to specific greenhouse gas emissions reduction targets, and are developing and publishing plans, strategies, and timelines to achieve those targets and to make progress toward other sustainability goals. They are developing standards and protocols by which progress can be tracked and assessed. And they are banding together in organizations for mutual support, consultation, and peer-to-peer engagement—ICLEI/Local Governments for Sustainability, C40 Cities, Urban Sustainability Directors Network, and others—that together constitute a vast stratum of activity humming beneath the high-level but sluggish international diplomatic processes.

We believe that this convergence of deliberation and localism in the sustainability movement is fortuitous. Precisely at the time when cities and communities are stepping up to chart their own ways forward into a warming and transforming world, deliberation is blossoming into a proven and potent means of harnessing the insights, commitment, buy-in, and action of ordinary people everywhere. As David Kahane notes in chapter 8, "a systemic deliberative democracy would support citizens in coming to grips with the wicked and socially complex character of climate change, so that they could shape wise and effective responses to the challenges."

Do Try This at Home

Growing citizen empowerment and greater fragmentation and polarization could result in political systems that make governing more difficult. [. . .] The digital age undermined many of the barriers that used to protect public authority, rendering governments

much less efficient or effective as the governed, or the public,
became better informed and increasingly demanding in their
expectations.

Klaus Schwab,

The Fourth Industrial Revolution

Significantly, the author of the above quote, Klaus Schwab, sees citizen empowerment as a bad thing. But then, he is the founder of the World Economic Forum, sponsor of the annual gathering of the world's political and business elites in Davos, Switzerland. We would argue the opposite point, that a certain constructive public resistance to being told what to do, even if that makes citizens more difficult to govern, is a good thing. People *should* become "better informed and increasingly demanding" in order to raise the odds of successfully confronting sustainability challenges. Deliberation is a useful way to promote that.

By now this essay may appear to be a hymn to deliberation. While we believe in its potential, we think it wise to guard against being too starry-eyed about it. If democracy is the worst form of government except for all others, deliberative democracy may be the worst form of democracy—except for all the other forms. That is, it's flawed. Localized communities or polities can easily go off the rails, like separate populations of organisms evolving in isolation. As Adolf Gundersen has noted, "purely local action will tend to be chauvinistic" (Gundersen 1995, 199). But surely in an Internet-connected world it should be more possible than ever to link our neighbourhoods, towns, and cities in "communities of regional communities," in Herman Daly and John Cobb's words (Daly and Cobb 1989, 176) and thereby to temper, to some extent, the excesses. Moreover, while we believe that deliberative democracy is probably better able to anticipate and cope with the changes in store due to climate change, we also believe that the community capacity cultivated where deliberation takes root will better enable those towns and cities to survive and prosper in a world where an increasingly deranged biosphere stresses, and possibly unravels, global social, political, and economic systems.

So, let a thousand deliberative flowers bloom. If there are expert practitioners available, by all means tap their knowledge and skills. Otherwise, go ahead— carefully!—anyway. The help needed to maximize the odds of success is available in many forms (see the website—www.albertaclimatedialogue.ca—for useful sources and organizations). And now is the time—among many people there is an ache for a system in which citizens take a larger role in managing their communities. Millions of us have become heartsick and cynical about the generally impoverished character of popular political discourse. We are appalled by the demagogic, sound bite-driven, corporate-funded, lowest-common-denominator election campaigns that typify politics in so many countries. That's why we are drawn to deliberation—we recognize that when people come together in a calm setting to think and talk about the issues that concern them collectively, interesting and positive things can happen: views shift and evolve, and people learn things. Sometimes they change their minds. Sometimes they cease to view those who disagree with them as Hell-spawn. Perhaps they become less susceptible to the kinds of one-dimensional and emotion-driven arguments that characterize contemporary public politics, and less willing to accept the outcomes delivered by the hidden machinery of backroom governance. While there might be less theatre in a world with more deliberation, can anyone doubt that our political lives would be better? The people, John Adams wrote, "must be taught to reverence themselves, instead of adoring their . . . generals, admirals, bishops, and statesmen" (cited in Rothman 2016). To put this in twenty-first-century terms, citizens and leaders need settings in which they will be more likely to reverence one another, and move from a parent-child relationship to one that is more equitable and complementary.

Three million years of hominid evolution have hard-wired us for functioning in small groups that are relatively "flat" in organizational terms. It remains to be seen whether this legacy equips us to confront and cope with complex, global problems requiring systemic thinking and large-scale, collective action by billions of people. But it is mainly our institutions that both channel and mitigate the good and bad tendencies built into our wiring as social primates, so we owe it to ourselves to refine our institutions, especially our governance institutions, in ways that align them with our evolutionary biology. Perhaps our long history of sitting around campfires together and talking about what's going on in the world around us, and what we ought to do about it next, can be put to good use.

References

Daly, Herman. 1991. *Steady-State Economics*. Washington, DC: Island Press.
Daly, Herman, and John Cobb. 1989. *For the Common Good: Redirecting the Economy toward Community, the Environment, and a Sustainable Future*. Boston: Beacon Press.

Galbraith, James K. 2014. *The End of Normal: The Great Crisis and the Future of Growth*. New York: Simon and Schuster.

Gordon, Robert J. 2012. "Is U.S. Economic Growth Over? Faltering Innovation Confronts the Six Headwinds." Working Paper 18315. Cambridge, MA: National Bureau of Economic Research. http://www.nber.org/papers/w18315.

Gundersen, Adolf. 1995. *The Environmental Promise of Democratic Deliberation*. Madison: University of Wisconsin Press.

Heinberg, Richard. 2011. *The End of Growth: Adapting to Our New Economic Reality*. Gabriola Island, BC: New Society Publishers.

Jackson, Tim. 2009. *Prosperity without Growth: Economics for a Finite Planet*. London: Earthscan.

Kemmis, Daniel. 1990. *Community and the Politics of Place*. Norman: University of Oklahoma Press.

Kreps, Bart Hawkins. 2016. "Can We Afford the Energy Demands of the 'Fourth Industrial Revolution'?" *Resilience*. http://www.resilience.org/stories/2016-01-21/can-we-afford-the-energy-demands-of-the-fourth-industrial-revolution-don-t-ask.

Leighninger, Matt. 2012. "Mapping Deliberative Civic Engagement." In *Democracy in Motion: Evaluating the Practice and Impact of Deliberative Civic Engagement*, edited by Tina Nabatchi, John Gastil, G. Michael Weiksner, and Matt Leighninger, 19–39. Oxford: Oxford University Press.

Nabatchi, Tina, and Matt Leighninger. 2015. *Public Participation for 21st Century Democracy*. San Francisco: Jossey-Bass.

Orr, David. 2013. "Governance in the Long Emergency." In Worldwatch Institute, *State of the World 2013: Is Sustainability Still Possible?* 279–91. Washington, DC: Island Press.

Peixoto, Tiago. 2012. "The Benefits of Citizen Engagement: A (Brief) Review of the Evidence." *Democracyspot*. https://democracyspot.net/2012/11/24/the-benefits-of-citizen-engagement-a-brief-review-of-the-evidence/.

Rothman, Joshua. 2016. "Shut Up and Sit Down: Why the Leadership Industry Rules," *The New Yorker*, February 29. http://www.newyorker.com/magazine/2016/02/29/our-dangerous-leadership-obsession.

Smil, Vaclav. 2010. *Energy Transitions: History, Requirements, Prospects*. Santa Barbara, CA: Praeger.

Spink, Peter K., and Nina J. Best. 2009. "Introduction: Local Democratic Governance, Poverty Reduction and Inequality: The Hybrid Character of Public Action." *IDS Bulletin* 40(6): 1–12. doi:10.1111/j.1759-5436.2009.00079.x.

Touchton, Michael, and Brian Wampler. 2014. "Improving Social Well-being through New Democratic Institutions." *Comparative Political Studies* 47(10): 1442–69. doi:10.1177/0010414013512601.

Victor, Peter, and Tim Jackson. 2015. "The Trouble with Growth." In Worldwatch
Institute, *State of the World 2015: Confronting Hidden Threats to Sustainability*,
37–49. Washington, DC: Island Press.

Worldwatch Institute. 2016. *State of the World 2016: Can a City Be Sustainable?*
Washington, DC: Island Press.

Contributors

Gwendolyn Blue is an associate professor in the Department of Geography at the University of Calgary. Her research interests include interpretive approaches to public engagement with environmental issues. She is currently working on a project that seeks to understand how corporate power influences dominant frames and public discussions about climate change. Ongoing work seeks to open public engagement with climate change to alternative perspectives, values, approaches, and world views.

Shelley Boulianne is an associate professor of sociology at MacEwan University. She completed her PhD in sociology at the University of Wisconsin-Madison. Her research focuses on media use and civic and political engagement, as well as survey research methodology. She has published in many journals, including *Political Communication, New Media and Society, Information, Communication and Society, Social Science Computer Review, Field Methods, Canadian Review of Sociology*, and the *International Journal of Public Opinion Research*.

Jacquie Dale, MM, MSc, CMC, is a public engagement specialist and partner at One World Inc., www.owi.ca. A multiple award winner for her work in deliberative dialogue, Jacquie is one of Canada's foremost engagement practitioners, having designed and facilitated well over five hundred sessions over the last twenty years. Increasingly, she works with organizations to develop the skills, strategies, tools, and resources to implement and evaluate effective public and patient engagement practices in their own work. Her reflections and lessons learned on deliberative dialogue as a practitioner have appeared in several published articles. In 2016, she co-authored "Framing and Power in Public Deliberation with Climate Change: Critical Reflections on the Role of Deliberative Practitioners" which was published in the *Journal of Public Deliberation*.

Susanna Haas Lyons, MA, is a civic engagement specialist. She designs participation strategies, facilitates complex meetings, and provides training for better conversations between the public and decision makers. Bridging online and face-to-face methods, she has worked for over fifteen years on some of North America's largest and most complex citizen and stakeholder engagement projects. Susanna teaches engagement skills for government, business, non-profits and at post-secondary institutions. Susanna is also a judge for IAP2 Canada's annual Core Values Awards, which recognize excellence and innovation in public participation. susannahaaslyons.com

Lorelei L. Hanson is an associate professor of environmental studies at Athabasca University and a fellow with the Energy Futures Lab, a social learning lab focused on identifying innovation pathways to disrupt and transition Alberta's energy system. Her research interests include energy transition, critical sustainability, food security, public dialogue on climate change, and environmental history, and her work can found in journals such as *Environmental Politics*, *The International Journal of Interdisciplinary Environmental Studies* and *Local Environment: The International Journal of Justice and Sustainability*.

David Kahane is a professor of political science at the University of Alberta, specializing in democratic theory and practice. From 2007 to 2010 he shepherded along the development of the project that became ABCD, and was ABCD's Principal Investigator and Project Director from 2010 to 2016. His current research focuses on citizen deliberation and systems change, including in relation to the nascent discipline of "systemic design."

Matt Leighninger is the Vice President for Public Engagement, and Director, of the Yankelovich Center, at Public Agenda. Public Agenda is a non-profit, non-partisan organization that helps diverse leaders and citizens navigate divisive, complex issues and work together to find solutions. Over the last twenty years, Matt has worked with public engagement efforts in over one hundred communities, in forty states and four Canadian provinces. Previously, Matt served as executive director of the Deliberative Democracy Consortium, an alliance of major organizations and leading scholars working in the field of deliberation and public participation. He has also assisted in the development of Participedia, the world's largest online repository of information on public engagement and authored two books, *The Next Form of Democracy*, and, with Tina Nabatchi, *Public Participation for 21st Century Democracy*.

Mary Pat MacKinnon, MPA, is a research practitioner, devoted to engaging citizens in public policy and program delivery. She writes and presents on the theory and practice of engagement and social change. Her career spans management positions, including as Vice-President, Hill + Knowlton Strategies; Partner, Ascentum, a digital and in-person engagement firm; Director, CPRN's Public Involvement Network; and Director, Government Affairs and Public Policy, Canadian Co-operative Association. Volunteer contributions include: Senior Fellow at the University of Ottawa's Graduate School of Public and International Affairs (2007–2015); Deliberative Democracy Consortium; Board Director, Friends of Canadian Broadcasting; jury member of the National Credit Union Award for Community Economic Development; and Chair, Low-Income Tax Relief Working Group, Ontario Fair Tax Commission.

John Parkins is a professor in the Department of Resource Economics and Environmental Sociology at the University of Alberta. His master's degree in rural sociology and PhD in sociology are from the University of Alberta. He joined the university after ten years working as a social scientist with Natural Resources Canada. His research and teaching cover a range of topics, including social perspectives on renewable energy transition, social impact assessment, and the politics of resource management in Alberta. His recent research on public engagement is published in the journal *Society and Natural Resources* and *Environmental Politics*.

Tom Prugh is a senior researcher at the Worldwatch Institute, co-directing five of the institute's *State of the World* reports and editing *World Watch* magazine. In 2013, he served as an editor and contributing writer for the Secretariat of the New Development Paradigm of the Royal Government of Bhutan. Before joining Worldwatch in 2002, he spent nine years at the US Energy Information Administration as a writer and manager. Tom has published in many journals and periodicals and is lead author of two books, including *The Local Politics of Global Sustainability* (with Robert Costanza and Herman Daly).

Geoff Salomons is a PhD candidate in the Department of Political Science at the University of Alberta. He also holds an MA in Political Science from the University of British Columbia. His research is broadly focused on the challenges of democratic governance of the environment and natural resources. His dissertation concerns the intergenerational issues surrounding non-renewable resource governance in Alberta. His master's thesis, analyzing the consequences of restricted public participation in Canadian environmental assessments, was published in the *Environmental Impact Assessment Review*.

Index

activism, 71–2
adaptation, 137, 138–39, 227
adaptive management, 34, 61
Alberta, Government of: and Alberta
    Energy Efficiency Choices, 49, 52,
    99–100; climate change policy, 93–95;
    and deliberative democracy, 69;
    land use strategy, 140; and public
    participation at regional level, 92–93,
    96; status as petro-state, 87–89; and
    Water in a Changing Climate, 55
Alberta Climate Dialogue (ABCD)
    (*see also* Alberta Energy Efficiency
    Choices (AEEC); Citizens' Panel
    on Edmonton's Energy Climate
    Challenges; City-Wide Food and
    Urban Agriculture Citizen Panels;
    Water in a Changing Climate):
    and activism, 72; assessment of its
    deliberation on climate change,
    60–61; collaboration in Citizens'
    Panel On Edmonton's Energy,
    152–7; collaboration with AEEA,
    157–58, 163; collaboration with
    CPI, 151–52, 164; collaboration
    with Oldman Watershed Council,
    159–60, 164; as collaborative project,

149, 150–51; context for launch of,
95–96; effectiveness of based on
eight measures, 203–8; focus on
values during deliberation, 61; and
framing of climate change, 76, 134,
138; goals, 197; history and legacy,
5–7; how it dealt with wicked issues,
60–62; how systems analysis could
improve work of, 216–19; impact on
field of deliberative democracy, 22,
208–9; and Indigenous participation,
75; lessons learned from projects,
101; policy context of deliberations,
172–73; and political context in
which climate change is viewed, 11;
political influence of, 69, 83; reasons
why social change was not addressed
by, 209–12; recruitment challenges
of, 126–29, 207, 217; role in Alberta
Energy Efficiency Choices, 48, 52;
role in Citizens' Panel on Edmonton's
Energy, 42, 46; role in City-Wide
Food and Agricultural Panels, 35;
role in Water in a Changing Climate,
54, 55, 58; scope of, 4; and social
diversity, 75, 76

on, 84–86; recommendations for framing, 143–44; research and evidence proving, 12–13; and social justice, 142, 143; thoughts on ABCD deliberations on, 60–61; and UN's Framework Convention Paris Agreement, 8; using systems theory to combat, 199–200; and values, 174; view of in Alberta, 89–90, 96–97; as wicked issue, 10–11, 74, 198–99

collaboration: in Alberta Climate Dialogue, 150–51; in Alberta Energy Efficiency Choices, 157–58, 162, 163; in Citizens' Panel on Edmonton's Energy and Climate Challenges, 152–57; in City-Wide Food and Urban Agriculture Citizen Panels, 151–52; in Water in a Changing Climate, 159–60, 162–63; defined, 147; factors supporting successful, 160–64; principles of, 147–48; used to address wicked issues, 148–49, 165

communication, 163–64

community empowerment, 203, 207

Conversation on Climate Justice, 209

creativity, 175, 179

Dale, Jacquie, 55, 141

Daly, Herman, 231

deliberative capacity, 202, 207

deliberative civic engagement (DCE), 232–34

deliberative democracy (*see also* Alberta Climate Dialogue (ABCD); framing issues; mini-publics; recruitment; social change; social learning; wicked issues): ABCD's effect on field of, 22, 208–9; advantages in addressing wicked problems, 3–4; and attitudinal diversity, 113–15, 119; as best method to address climate change with, 74–78; characteristics of, 67–68; Citizens' Panel on Edmonton's Energy view

of, 47–48; claim for citizens' right to, 169–70; and creativity, 175, 179; critical views of, 77–78, 133–34, 201; as deliberative civic engagement, 232–34; and demographic representation, 111–13; described, 17; divergence-convergence of, 216–17; effect of use of systems analysis in, 20–21, 218–19; effect on systems theory, 219–20; eight areas of divergence and debate in, 68–73; engagement of social change, 200–3, 209–15; and environmental regulation, 91–93; further resources on, 235–36; history, 17, 83; ideas for improving, 77–78; importance of values to, 61, 68; incentives for taking part in, 117; involvement of non-experts in, 226–27; its ability to change participants, 203, 207–8, 234–36; and Oldman Watershed Council, 140; one-off events v. systems, 69–70; and ownership of project, 188–91; possibilities of, 229, 235–36; professionalization of, 19–20, 72–73; pros and cons of, 19–20; recommendations for framing climate change in, 143–44; recruitment strategies and approaches for, 109–17; scholarly work on, 17–18; significance of framing issues for, 134–36; and social diversity, 75–76; and social learning, 61–62; as sustained engagement, 233–34; technology and, 52–54, 180, 189; terminology of, 18–19; and time issues, 164, 210–11, 218; and trust, 228–29; and use of collaboration, 148–49

Deliberative Democracy Consortium (DDC), 5

deliberative society, 70